The Palgrave Lacan Series

Jacques Lacan is one of the most important and influential thinkers of the 20th century. The reach of this influence continues to grow as we settle into the 21st century, the resonance of Lacan's thought arguably only beginning now to be properly felt, both in terms of its application to clinical matters and in its application to a range of human activities and interests. The Palgrave Lacan Series is a book series for the best new writing in the Lacanian field, giving voice to the leading writers of a new generation of Lacanian thought. The series will comprise original monographs and thematic, multi-authored collections. The books in the series will explore aspects of Lacan's theory from new perspectives and with original insights. There will be books focused on particular areas of or issues in clinical work. There will be books focused on applying Lacanian theory to areas and issues beyond the clinic, to matters of society, politics, the arts and culture. Each book, whatever its particular concern, will work to expand our understanding of Lacan's theory and its value in the 21st century.

More information about this series at
http://www.palgrave.com/gp/series/15116

Svitlana Matviyenko · Judith Roof
Editors

Lacan and the Posthuman

palgrave
macmillan

Editors
Svitlana Matviyenko
School of Communication
Simon Fraser University
Burnaby, BC, Canada

Judith Roof
Rice University
Houston, TX, USA

The Palgrave Lacan Series
ISBN 978-3-030-09466-9 ISBN 978-3-319-76327-9 (eBook)
https://doi.org/10.1007/978-3-319-76327-9

Cover illustration: Peter Crowther/Alamy Stock Photo

Printed on acid-free paper

This Palgrave Macmillan imprint is published by the registered company Springer International Publishing AG part of Springer Nature
The registered company address is: Gewerbestrasse 11, 6330 Cham, Switzerland

Acknowledgements

We would like to thank the Palgrave Lacan Series editors Derek Hook and Calum Neill for their support of this volume as part of the series and for designing a Lacanian niche in academic publishing on the first place. We would also like to thank Joanna O'Neill and Grace Jackson for their expert guidance through the process.

Our meeting during 2011 MLA Convention in Los Angeles was facilitated by our friend Benjamin Kozicki, and it led to an inspiring conversation about Lacan and posthumanism, in the midst of which we understood we wanted to co-edit a book. We thank those scholars who responded to our initial call and shared this project's agenda. For a variety of reasons, the preparation of this volume took significantly longer than expected, so we are grateful to our contributors for their commitment to this project and for their patience. We appreciated the suggestions of the anonymous reader, who enthusiastically engaged with our proposal, which helped us to locate a gravity centre of this collection: the notion of "post" in posthumanism. We thank Judith's assistants Brooke Clark and Hannah Biggs for their help with copyediting. And finally, we thank Ellie Ragland for her Lacanian inspirations.

Contents

Editors and Contributors

About the Editors

Svitlana Matviyenko is Assistant Professor of Critical Media Analysis in the School of Communication at Simon Fraser University, Canada. She writes about history of science, cybernetics and psychoanalysis; the networking drive and user complicity; practices of resistance and mobilization; legacies of the Soviet techno-politics, including the Chernobyl catastrophe; information and cyber-war. With Paul D. Miller, she is co-editor of *The Imaginary App* (MIT Press, 2014) and a co-author, with Nick Dyer-Witheford, of *Cyber-War and Revolution* (forthcoming with Minnesota in 2019).

Judith Roof is Professor and William Shakespeare Chair in English at Rice University. She is the author of seven monographs, most recently *What Gender Is, What Gender Does* (Minnesota 2016) and *The Comic Event: Comedic Performance from the 1950s to the Present* (Bloomsbury 2018). She is also an attorney.

Contributors

Louis Armand is the director of the International Arts Centre Prague and is the author, among other works, of *Incendiary Devices: Discourses of the Other* (1993), *Literate Technologies: Language, Cognition, Technicity* (2006) and *Videology* (2015). www.louis-armand.com.

Nancy Gillespie is an analyst in private practice. She is a member of the Lacanian Compass, an affiliated group of the New Lacanian School. She is co-editor of the *Lacanian Compass Express*, has done English editing for several Lacanian journals and is guest editing a section of *The Lacanian Review 5: The Delights of the Ego*. She has a PhD in literature and psychoanalysis from the University of Sussex in the UK and has published work on experimental writing and Lacanian analysis.

John Johnston is Professor in the Department of English at Emory University. He is the author of *The Allure of Machinic Life* (MIT, 2008); *Information Multiplicity: American Fiction in the Age of Media Saturation* (John Hopkins UP, 1998); and *Carnival of Repetition: William Gaddis' The Recognitions and Postmodern Theory* (University of Pennsylvania Press, 1990). Johnston is an editor and translator of *Literature, Media, Information System: Essays by Friedrich A. Kittler* (Routlege, 1997).

Timothy Morton is Rita Shea Guffey Chair in English at Rice University. He gave the Wellek Lectures in Theory in 2014 and has collaborated with Björk, Haim Steinbach and Olafur Eliasson. He is the author of *Dark Ecology: For a Logic of Future Coexistence* (Columbia, 2016), *Nothing: Three Inquiries in Buddhism* (Chicago, 2015), *Hyperobjects: Philosophy and Ecology after the End of the World* (Minnesota, 2013), *Realist Magic: Objects, Ontology, Causality* (Open Humanities, 2013), *The Ecological Thought* (Harvard, 2010), *Ecology without Nature* (Harvard, 2007), eight other books and 160 essays on philosophy, ecology, literature, music, art, architecture, design and food. Blog: http://www.ecologywithoutnature.blogspot.com.

Allan Pero is Associate Professor of English and Core Faculty at the Centre for the Study of Theory and Criticism at the University of Western Ontario, London, Canada. He teaches modern literature, literary theory and modern drama. He is co-editor (with Gyllian Phillips) and contributor to a volume entitled *The Many Façades of Edith Sitwell* (Spring 2017), with U Press of Florida. Recent publications include articles in the *Wordhoard, The Journal of Wyndham Lewis Studies, The Virginia Woolf Miscellany, Modernism/modernity, Katherine Mansfield Studies* and a forthcoming chapter on Ford's *The Good Soldier*. He is currently working on *An Encyclopedia of Cultural Theory*, with Dr. Kel Pero for the University of Toronto Press, and a book-length project on Camp and Modernism.

Scott Wilson teaches Media and Psychoanalysis at Kingston University, London. He is a regular contributor to *The Lacanian Review Online* edited by Marie-Hélène Brousse and France Jaigu. He is the author of *Stop Making Sense: Music from the Perspective of the Real* (Karnac, 2015). He is currently working on a short book about Lacan and Alan Turing for Bloomsbury.

Ben Woodard is a post doc researcher at Leuphana University in Luneburg, Germany. He writes about the relation between non-eliminative naturalism (Schelling, Iain Grant), idealism (Schelling, Bradley) and pragmatism (Peirce, James, etc), as well as on nihilism, pessimism, weird fiction and horror film. Woodard is the author of first book *Slime Dynamics: Generation, Mutation and the Creep of Life* (ZeroBooks, 2012) and *On an Ungrounded Earth: Towards a New Geophilosophy* (Punctum, 2013).

Colin Wright is Associate Professor of Critical Theory in the Department of Culture Film & Media at the University of Nottingham, UK. He is the Director of the MA in Critical Theory and Cultural Studies there. His research interests include Lacanian psychoanalysis, French critical theory and postcolonial theory. He is the author of *Perversion Now!* (co-edited with Diana Cain, 2017), *Badiou in Jamaica: The Politics of Conflict* (2013), *Psychoanalysis* (2008), *Post-Conflict Cultures: Rituals of Representation* (co-edited with Cristina Demaria, 2006), and *Philosophy, Rhetoric, Ideology: Towards a Sophistic Democracy* (2006). His current book project is entitled *Toxic Positivity: A Lacanian Critique of the Happiness and Wellbeing Agenda*. He is also a Lacanian psychoanalyst in private practice in Nottingham in the UK.

1

Introduction

Svitlana Matviyenko and Judith Roof

Loss, Lack

When Ihab Hassan introduced the term "posthumanism" in 1977, it marked a state of powerlessness and disorientation in face of rapidly changing humanities.[1] "We need first to understand that the human form—including human desire and all its external representations—may be changing radically, and thus must be re-visioned", he urged. "We need to understand that five hundred years of humanism may be coming to an end, as humanism transforms itself into something that we must helplessly call posthumanism".[2] Characterized as the crisis of

S. Matviyenko (✉)
School of Communication, Faculty of Communication,
Art and Technology, Simon Fraser University,
Burnaby, BC, Canada
e-mail: smatviye@sfu.ca

J. Roof
Rice University, Houston, TX, USA
e-mail: roof@rice.edu

S. Matviyenko and J. Roof (eds.), *Lacan and the Posthuman*, The Palgrave Lacan Series,
https://doi.org/10.1007/978-3-319-76327-9_1

humanism's vocabulary forty years ago by Hassan, posthumanism presaged both the fear and anticipation of a shift in some of humanism's more dubious dichotomies, or even its practice of binarizing conceptual terrain itself. Nature/culture, subject/object, human/machine, human/animal/animal: all seemed to be the products of a humanist tendency to reduce and simplify on behalf of a conception of world order that made the position of human as arbiter clearly apparent.

Even if pushing past humanist assumptions ultimately resulted in the loss of a worldview in which the human was both central arbiter and beneficiary, it is only through *loss*, as psychoanalysis understands the term, that we experience *lack. Lack*, or a *manque-à-être*, not only generates desire, but also reflects the (pre) ontological status of being itself as constituted by "the gap of the unconscious".[3] The loss of human centrality represented by the shift away from humanist ideals, thus, repeats a more profound, traumatic lack. Conjuring a philosophical system in which humans imagine themselves as subject to rather than authors of a worldview might be a sobering encounter, were it not for the variety of posthumanist efforts to come to terms with, reimagine, resituate, and even recommend the effects of human decentralization in circumstances conceived to arise from a much broader and diverse set of possibilities.

Negativity, Subject

"The subject is no one"; "[i]t is decomposed, in pieces",[4] Lacan suggests and he warns us: "When one speaks of subjectivity, the problem is not to turn the subject into an entity".[5] Conceiving of the human subject as already the effect of multiple processes, fictions, and solutions to the ineffable, psychoanalysis has been perhaps the least comfortable with humanist outlook, especially in so far as the centrality of the human subject in humanist worldviews elaborates a more certain and stable subject than the split and contingent beings of quotidian existence. But psychoanalysis, as many of the essays in this volume demonstrate, has also already offered a way of thinking that can explore the ways cultural texts—film, literature, philosophy—have enjoyed, normalized, and/or foreclosed the dis-ease that derives from humanist suppositions of subjective control and stability. Psychoanalytically speaking, it is not enough to say "we have never been human". There is always the lurking issue of what being speaks that can make such a declaration.

Although the shadowy presence of the subject's illusion is imbedded in culture and society as well as our own sense of individual subjectivity, both humanism and posthumanism devise sites for the persistence of the subject, even in its apparent denial. Despite themselves, centuries of thinkers working within humanist assumptions have encountered the lack at the core of species-beings' ex-sistence. Humanism's response has been to erect an ethical system centering the human as custodial to all beings—a system that had the effect of orthopedically concealing subjective lack. Acknowledging the human subject as a continuous part of the system instead of as its governing exception—an insight of which psychoanalysis was already well aware—posthumanism provides another perspective on what psychoanalysis already knew about humanism's subjects. Such posthumanist thinkers as N. Katherine Hayles have characterized this insight about the human subject by articulating the affirmative "we have always been posthuman", instead of confronting the sheer *negativity* of the lack that dwells at the basis of Lacan's notion of the subject.[6]

Temporality, Future

It is evident in contemporary thought that the subject's lack-in-being does not reside solely in the past, but also inhabits the future. *Time travel* has returned as a trending topic in sci-fi and scholarly inquires,[7] as we now ponder the possibilities of a new form of determinism coming from the future. Dispensing with temporal linearity, some thinkers pronounce the future as already stolen or alienated from us,[8] already "written" at the cost of our well-being or freedom. In Lacan's work, the determinative sense of futurality is expressed in the cybernetic terms: "the letter always arrives to its destination", provided that its destination is where it arrives. This makes psychoanalytic thought well equipped to encounter and confront the complexities of the invasive pre-emption and premediation imposed by the current techno-political regimes that have been drawing the attention of scholars in media studies, political-economic studies, and sociology for several decades. When the future emerges in the present, the sense of "leaving behind" or "overcoming" or "losing" or "transgressing" in regard to what may come next suggests that "posthumanism" is a misnomer.

In this volume, essayists trouble posthumanism's sense of "post". The "post" in posthumanism is not the indicator of either the serial displacement of one philosophy by another or some species of intellectual or perceptual progress or even a logical ordering. Although it seems to supply its own self-temporalization, the posthuman, as it always has, undergirds, coexists beside and within, and plies after a humanism that itself persists. The "post" in posthuman is as much scalar and perspectival as it is a polysemous temporal configuration. Seeming to occupy a point of view from a grander scale than the humanist so as to correct humanism's myopic mistakes, the "post" of posthumanism imagines a more perceptive locus for the subject it thinks it has left behind (or below or on a different scale) relegated to an egalitarian status with all else. This illusion of an adjusted scalar perception is partly an effect of an increasing ability to discern the operations of larger systems and networks within which the human, among all else, is ensconced as a contributing part. This attention to systemic complexity is less interested in human responsibility and conceptual categories defined in language, but in a host of material relations imagined to be real. Finally, the "post" of posthuman is the effect of an imaginative leap made by human subjects to transcend the human subject by observing the human's constitutional weakness from a point simultaneously distant and suddenly endowed with the insight of *le-sujet-suppose-savoir*.

Nonhuman, Imaginary

The posthuman shares its "nonhuman" turn with other non-anthropocentric humanistic approaches. Generated by the rise in technologies (digital computers, biotechnological engineering, nanotechnologies, artificial intelligence) that in their operation seem to decenter the human as the site of apprehension and linguistic reifications, these approaches appear to offer models wherein all (eco-)systems operate like technologies instead of, as perhaps formerly, all complying with humanist perceptions of hierarchically arranged systems. Most of these approaches—the new materialism, cognitive science approaches to aesthetics, actor-network theory, affect theory, object-oriented ontology, systems theory, and the inhuman, discussed notably by Jean-François Lyotard in his study, *The Inhuman*[9]— appear to subordinate the human subject to broader, more inclusive conditions and systems through which the human becomes one element among many interacting agents, causalities, and quantities. Still, however, the site from which these systems can be imagined does not really disappear.

The analogy of the subject itself as system that already appears in Lacan's *oeuvre* suggests that these enlargements of technological aegis not only mask the still-centered human subject, but also partially model the relations among a species of consciousness and systematicity that forms the human subject itself, in so far as the human subject both models and organizes the relations among biological and physical processes and environments, other entities engaged in the same processes, and the balancing acts of survival. And just because humans envision a broader spectrum of operations in which their will does not govern does not mean that the human subject itself either disappears or ceases to imagine its place in systems—or that human subject can ever imagine anything other than projected anthropomorphism.

Anthropocene, Human

At the time of this volume's completion, the term "Anthropocene" that was introduced in public debates in the early 2000s by atmospheric chemist Paul J. Crutzen has not received an official approval by the International Commission on Stratigraphy and the International Union of Geological Sciences as a legitimate geological epoch. If accepted, it will destabilize too much: not only it will demand a new concept of history—biological or geological,[10] it will also pose significant demands for a different mode of political thought, different bases for the economy, and a largely revised notion of ethics.[11] It should, thus, not surprise us that the symbolic recognition of a man-made catastrophe is delayed, despite the fact that we continue witnessing the returns of the real on the body of the planet. This delay, however, is not merely surprising, it is shocking, because, as Crutzen notes, "[m]ankind's growing influence on the environment was recognized as long ago as 1873, when the Italian geologist Antonio Stoppani spoke about a 'new telluric force which in power and universality may be compared to the greater forces of earth,' referring to the 'anthropozoic era'".[12] Vladimir Vernadsky followed Stoppani with similar warnings about the growing impact of humans on their surroundings. Introducing the influential notion "noösphere", which Vernadsky regarded as "a new geological phenomenon on our planet", he proclaimed that "in it, for the first time, man becomes a large-scale geological force".[13] Other Soviet scholars raised the issue in the 1960s amid a chorus of concerned voices.

Is it scandalous that so long history is hidden behind the "sudden" emergence of Anthropocene, or is it typical misrecognition that exposes the complexity and contradictions in our relation to knowledge? Lacanian psychoanalysis has observed that to know is never enough for action. Too often, the obstacle to confronting knowledge comes from within and is anchored in the very ontology of the subject itself. Lacan spelled this notion of shameful ontology "*hontology*", where French "*honte*" stands for "shame". Apparently, it is less painful for the subject to turn away from a deadly disaster, even at the cost of death, than to survive the disgrace of admitting that we knew what we were doing, that we knew about the potential consequences of our actions, and that we were both passively and actively complicit in the wrongdoings of others.

The real emergence of the Anthropocene—the era in which the direction and organization of life and geological processes have been and will continue to be affected by human action—belongs to the past. The awareness that human action could potentially alter the direction of biological and geological processes has changed both the vectors of events as well as notions of causality and the predictability of the future. This new future torqued by human action generates a new sense of mortality, in which humans are now bound in death with all other species on the planet, creating a novel sense of commonality on what might be the threshold of extinction. The entwined perceptions of language and matter, humanism and posthumanism face a potentially truncated future unless ethics and perceptions work together. The Anthropocene is a traversed fantasy of posthumanism and the ultimate encounter with the real as the fragility of life.

Prometheus, Oedipus

Hassan evokes the myth of Prometheus, who "is himself the figure of a flawed consciousness struggling to transcend [...] to express the sense of transgressing boundaries and offending tradition Hassan perceives in emergent posthumanism. The most relevant aspects of the Promethean dialectic to posthumanism itself are Imagination and Science, Myth and Technology, Earth and Sky".[14] In his exposition of the Aeschylean theology, Harold Bloom notes that "Freud called *Oedipus* an 'immoral play,' since the gods ordained incest and parricide. Oedipus therefore participates in our universal unconscious sense of guilt, but on this reading so do the gods". Indeed, the Other experiences guilt, and therefore, the Other lacks just like the subject does. Betrayed by the impotent Other, Bloom confesses, "I sometimes wish that Freud had turned to Aeschylus instead, and given us the Prometheus complex rather than the Oedipus complex".[15] But what would it be?

As a way of approaching the confluence of pride and guilt, mythology, so much loved by both Freud and Lacan, also offers an end to Prometheus's story. According to Hesiod's "Theogony", Prometheus disobeys, steals fire, is punished, and then, eventually, freed by Heracles. But the details of this last part survived only in fragments and are blurry enough to inspire too many poetic variations of events. A century ago, as Vernadsky was formulating his theory of a "new geological force", Franz Kafka commented on the conflicting endings of the great Promethean myth by suggesting that the more one thinks about it, the less grandiosity one finds in the actions of this Titan. Where is the creator of humanity, for example, from the point of view of deep history, if "his treachery was forgotten in the course of thousands of years, forgotten by the gods, the eagles, forgotten by himself"? In the end, Kafka suggests, "everyone grew weary of the meaningless affair. The gods grew weary, the eagles grew weary, the wound closed wearily". And he closes his speculation by pointing to the presence of the fossil of the real at the core of the Prometheus complex: "There remains the inexplicable mass of rock. The legend tried to explain the inexplicable. As it came out of a substratum of truth it had in turn to end in the inexplicable".[16] The true grandiosity in deep time belongs to the slow mineral life of the thing. Who is Prometheus, or any of us, in its presence?

For Mladen Dolar, however, nothing expresses the subject's condition better than the story of Oedipus. Borrowing Alenka Zuzančič's dark formula of comedy, "Not only we are not infinite, we are not even finite",[17] Dolar depicts Oedipus as wriggling out of it all: "Not only have I committed a horrible crime, I cannot even be guilty of it".[18] This helps to articulate the complexity of collective responsibility in the Anthropocene. But does the impossibility to identify one's particular part in the collective crime make easier or more difficult for this subject to live with guilt? Not only humans cannot rise to the perfection of gods by transgressing death, as it has been already discovered by the end Renaissance, they are not even protected in death from further entering some unthinkable live assemblages in the form of organic matter. In an era of enfolded humanism/posthumanism, what conceptions might guide the subject in all of its frailty and presumption toward a better understanding of the relations among the world, its beings, and the human we can never stop being?

Lacan, Posthumanism

For those who consider Jacques Lacan's psychoanalytic philosophies, one source of such guiding insights is the work of Lacan himself, especially as psychoanalysis already situates the subject as lacking. What might a Lacanian perspective on issues of loss, lack, the unconscious, and the subject's perpetual fragmentation bring to an understanding of the motivations and queries raised by posthumanism? How might Lacanian psychoanalysis situate and account for the displacements of the kind of subject posthumanism and its effects produce and describe? As responses to posthumanism's changes in perspective and enlarged range of potential agencies, the essays appearing in this volume explore the conjunctions, intersections, and collisions of Lacanian thinking and conceptions of posthumanism as these manifest in political thought, law, style, film, literature, performance art, hysteria, and feeling.

The first three essays consider the functions of Lacan's Borromean knot and the sinthome that ties it together. Reading Lacan via Marx, Louis Armand's "The Obscene Object of Post/humanism" shows how the category of the posthuman is something more than a surplus produced by a system of technologization bound to political economy: it shares the "primordial" evanescence of the commodity and of spectacle, as a dream standing before the mirror of mechanical reason. For Lacan, Marx's conceptions of the relation between technology and the subject imply that consciousness relates to a cybernetic materialism which, by way of the figure of Joyce in *Seminar XXIII*, leads to a theory combining topology, textuality, and the sinthome (or "synthhomme"). Judith Roof's essay, "From Law to Code: Posthumanism as Sinthome", locates "posthumanism" as a symptom of a broader shift from a law that operates metaphorically, as exemplified by Lacan's "Law-of-the-Name-of-the-Father", to a code operating metonymically represented by such materialist concepts as DNA and forensic evidence. Exchanging the function of the name for metonymies of matter suggests that the character of the "sinthome"—a fourth imaginary loop that Lacan theorizes tying together the symbolic, imaginary, and real of his Borromean formulation—has also morphed from a basis in language to the imaginary of materialism misunderstood as real and foundational. In "A Corporal Radioscopy: Lacan, the Baroque, and the Posthuman" Allan Pero argues that Lacan's Borromean knot is not an illustration, but a topological machine, working to re-conceptualize *jouissance*. The sinthome's particular "corporal radioscopy" reveals a radical substance: a baroque, machine of individuation producing a posthuman *jouissance*, independent of the phallic function. For Lacan, the soul is a machine that thinks; the baroque thus stages a thinking of the *jouissance* of being.

Essays by Colin Wright and Ben Woodard read films through a Lacanian lens for insights about posthumanism. Wright's "Lacan's Theory of Cybernetic Causality: Repetition and the Unconscious in *Source Code*" outlines how Lacan's early engagement with cybernetics and game theory informed his psychoanalytic theory of causality. Although at times, Lacan seems close to a posthumanism in his emphasis on the machinic or combinatorial aspects of the psyche, Wright's Lacanian interpretation of Duncan Jones' sci-fi film *Source Code* (2011) demonstrates that Lacan's thought stresses the temporality of desire and the act in human subjectivity, rather than machinic repetition. Ben Woodard offers an analysis of the numerous bug-human-machine hybrids in David Cronenberg's films as a way to explore Lacan's purported dismissiveness of the materiality of information in "A Fly in the Appointment: or Posthuman-Insectoid-Cyberfeminist-Materiality?" Engaging with the work of Lorenzo Chiesa, Jussi Parikka, Jakob von Uexkull, and Eugene Thacker, Wright's essay explores the materialization of information enabled by the technological affinities of insect anatomy (such as antennae) to technologies of communication, robotics, and military technologies.

Turning to issues of writing, desire, and drive, the next four essays explore the ways psychoanalyses' attention to issues of desire undercuts some of posthumanism's perceptual masquerade. Svitlana Matviyenko's "Graphocentrism in Psychoanalysis" is focused on the notion of "writing" and explores *graphocentrism* (as opposed to the alleged *logo-* or *phonocentrism*) that defines psychoanalysis. In "Lacan's Drive and Genetic Posthumans: The Example of Margaret Atwood's Oryx and Crake", John Johnston argues for the relevance of biotechnology and genetic engineering in defining the posthuman body and the desire it produces, showing how the relationship between the two informs a reading of Margaret Atwood's mapping of desire in *Oryx and Crake*. The absence of desire in the genetically engineered posthuman Crakers calls for a deeper analysis of desire itself, found in Lacan's differentiation of "drive" from desire and suggesting finally that the Oryx, the third major character of the novel, who occupies a fluid and mobile position between drive and desire, intimates another form of the posthuman.

Clarifying that desire, for the speaking being, has always been posthuman because it is the outcome of our interaction with the "machine" of language, Nancy Gillespie's "Posthuman Desire: The One-All-Alone in *Her*, *Ex Machina*, and *Lars and the Real Girl*" explores how this posthuman desire is, nevertheless, changing in our era. Using Jacques-Alain Miller's evolving concept of the "one-all-alone", and Zygmunt Bauman's notion of "liquid love" in late capitalism, Gillespie discusses how these changes are depicted in three contemporary films, *Her*, *Ex Machina*, and *Lars and the Real Girl*, which have "uncanny parallels" with our historical moment. In "Merzbow and the Noise of Object Oriented Perversion", Scott Wilson looks at how the artist Masami Akita seeks to restore lost *jouissance* to this desolate world through erotic art practices that evidence instances of what Lacan calls *père-version* and attempts to sustain a trace of non-digital, perverse subjectivity in a world dominated by objects filled with "intelligence" and techno-affectivity that orient subjects not toward another desire but toward pre-programmed demands which by-pass desire.

The final essay raises questions about who we are and what we can feel as humans in the posthuman. In "Melancholy Objects: If Stones Were Lacanian", Tim Morton suggests that melancholy, as an object-like presence, implies coexistence by definition. But that melancholy implies nothing in particular about subjectivity distinguishes it from other affective states, demonstrating that humans have more in common with nonhumans than they might suspect. Melancholy opens up a necessarily traumatic coexistence between humans and nonhumans, supplying a better reading of the death drive than one that presents it as a simple existential human drama.

Notes

1. Suzanne Dow and Colin Wright, "Introduction: Toward a Psychoanalytic Reading of the Posthuman," *Paragraph 33.3* (2010): 299.

2. Ihab Hassan, "Prometheus as Performer: Towards a Posthumanist Culture?," *Performance in Postmodern Culture*, eds. Michael Benamou and Charles Caramella (Madison: Coda Press, 1977), 212.
3. Jacques Lacan, *Seminar XI*, 29.
4. Jacques Lacan, *Seminar II*, 54.
5. Ibid., 53.
6. Claire Colebrook, *Death of the PostHuman: Essays on Extnction*, Vol. 1 (Ann Arbor: The Open Humanities Press and the University of Michigan Library, 2014).
7. For instance, see William Gibson's *The Peripheral* (2014) and James Gleick's *Time Travel: A History* (New York: Pantheon Books, 2016).
8. Franko "Bifo" Berardi, *Futurability—The Age of Impotence and the Horizon of Possibility* (London, Verso, 2017) and *After the Future*, eds. Gary Genosko and Nicholas Thoburn (Edinburgh: AK Press, 2011); as well as Zygmunt Bauman's "sharp dissection of our contemporary romance with the past," *Retrotipia* (London: Polity, 2017).
9. Jean-Francois Lyotard, *The Inhuman: Reflections on Time*, trans. Geoffrey Bennington and Rachel Bowlby (Stanford: Stanford University Press, 1991).
10. In the lecture "The Anthropocene, A New History?" (2015), Catherine Malabou discussed two new visions of history: a "biological history" introduced by Daniel L. Smail in the book *On Deep History and the Brain* (2007) and a "geological history" proposed by Dipesh Chakrabarty in the essay "The Climate of History: Four Theses" (2008).
11. Joanna Zylinska, *Minimal Ethics for the Anthropocene* (Ann Arbor: The Open Humanities Press and the University of Michigan Library, 2014).
12. Paul J. Critzen, "Geology of Mankind," *Nature* 415.3 (2002): 23.
13. Vladimir Vernadsky, *The Noösphere* (1943).
14. Ihab Hassan, 207.
15. Harold Bloom, *Bloom's Guides: Oedipus Rex* (New York: Chelsea Press, 2007), 8.
16. Franz Kafka, "The Complete Stories," ed. Nahum N. Glatzer (New York: Schocken Book, Inc., 1971).

17. Alenka Zuzančič, *The Odd One In: On Comedy* (Cambridge: MIT Press, 2008), 53.
18. Mladen Dolar, "Not Even. The Politics of Oedipus," *Jacques Lacan: Between Psychoanalysis and Philosophy*, eds. Samo Tomšič and Andreja Zevnik (New York: Routledge, 2016), 63.

2

The Obscene Object of Post/Humanism

Louis Armand

If we are inclined to suppose that a certain mode of questioning, of reflection and of self-consciousness should in itself characterise the meaning of the term Humanism, then the advent of the post-human would seemingly presuppose the resolution—independent of man (and independent of any Sartrean dialectic of reason)—of *what Humanism is*. On the other hand, it could also appear possible to say that an understanding of the human (of "the more or less presupposed unity of a [human] subject"),[1] by way of a *Humanism*, has always implied, however paradoxical this might sound, a stance necessarily coterminous with what today is called post-humanism—the apparent objectification of subjective experience, the world as prosthesis, artificial intelligence, virtual reality, *degré zéro des choses, degré zéro de la personne* … and so on—but without necessarily sharing aspirations to any kind of *transcendence or negation* of the human. It is a simple enough contention that in the

L. Armand (✉)
Centre for Critical and Cultural Theory, UALK,
Charles University, Prague, Czech Republic

S. Matviyenko and J. Roof (eds.), *Lacan and the Posthuman*, The Palgrave Lacan Series,
https://doi.org/10.1007/978-3-319-76327-9_2

discourse of post-humanism neither the human nor Humanism is effectively transcended or negated. Rather, something entirely else is at work, and this "something," this *thing*, is nothing less than the recursive *form* of Humanism's self-reflexive appeal to a condition of immanence. The pivotal moments of this thought are arguably Freud's discovery of the unconscious, the exploration in modernist writing of the conjunction between a theory of the unconscious and discoursivity, and Lacan's refiguring of the Freudian idea in his long-running seminar—in particular the 1954 series on cybernetics, and the 1976 series on topological chains and the work of James Joyce.

The Textual Thing

Lacan's engagement with cybernetics, and his treatment of the status of the *thing* in Freud ("A Materialist Definition of the Phenomenon of Consciousness," "Homeostasis and Insistence," "Freud, Hegel and the Machine," "The Circuit"), is conspicuous for the way it eschews any easy gesture of transcendence of the category of the human or of Humanism in approaching a definition of consciousness not seated in the *I* of Cartesian subjectivity, but instead positing an *antecedent subjective condition* of an apparently paradoxical *post*-human/ism. That is to say, the positing of a *post*-human/ism as the condition for any human/ism, by which subjectivity is grasped as an *operation*, or set of operations; in short, we might say, a *poiēsis*. Not simply an inversion of the classical dichotomy of *technē* and *poiēses*, but a thinking anew of a techno-poetic foundation of subjectivity: *techno* to the extent that subjectivity always involves a logic of prosthesis and is aim-orientated, *poetic* in the sense that subjectivity is tropic.

Here, Lacan echoes Marx in the *Grundrisse der Kritik der Politischen Ökonomie*, where he argues to the effect that the category of the "individual" is not abolished or placed in jeopardy by the experience of technology, but rather that technological "alienation" is in fact the necessary condition for any thought of the "individual" as such: the "machine" is identified by Marx as a mode of revealing of "man's" essential being. Alienation is in fact the condition (and meaning) of subjectivity. In this way, we see the individual/human as proceeding from an experience of the technological—the implication for Lacan being that Humanism can thus be thought in terms of a cybernetic materialism, which unfolds across his subsequent examinations by way of Joyce in *Seminar XXIII* of topology, textuality and the *sinthome* (or "synth-homme," we might say; the coincidence of a *synthetic humanism* and the Cartesian "subject" as symptom).[2]

Very early in his project, the active intransitive subjectivity of Descartes—"I think..."—is transfigured by Lacan into a transitive yet *subjectless* "I is thought ...," a movement predicated upon what we might call an *anteriority* of the subject that cannot however be represented as such, but which articulates the subject in language as, so to speak, its symptom, pointing the way to Lacan's treatment of the "sinthome" (and of the subjectlessness of the subject) in Joyce—a concern again and again reprised throughout the latter's work, most notably *Ulysses* and *Finnegans Wake*, which Lacan engages at length, but also, and perhaps more tellingly, in Joyce's earliest prose work, which Lacan largely overlooks. It is here that the problem of subjectivity is figured most explicitly in terms of, and in tension with, the assumptions of an agency that articulates itself, whose status is bound to that of the materiality of articulation, to the missed encounter between the Symbolic and Real, so that increasingly in the later seminars the former cedes priority in Lacan's investigations to the latter. Where the *Écrits* begins with a structuralist conception of the *letter*, Lacan's seminar on Joyce is a loosely coded examination of the status of *writing*—both as a species of thing and as an intransitive act cognate with terms linked to *being* and to a certain conception of agency, in consideration of the question: What is it that writes in Joyce?

Between *Dubliners* and *Finnegans Wake*, "la piste de Joyce" configures itself as a series of writing *mechanisms* which are described recursively, as a textual "body" that writes or signifies upon its own surface yet which defies analysis ("le sinthome de Joyce," Lacan allows, "est inanalysable").[3] This body assumes the function of both an agent (something that acts) and a *thing*. Contrary to certain principles that vest the act of writing uniquely in a human "subject," the Joycean "thing" adopts the position of a writing that returns out of what Slavoj Žižek refers to as the *obscene* realm of the Real and inscribes subjectivity as a fantastical "interface." Žižek's terminology, here, points to the *radically* Cartesian formulation that underwrites Freud's thinking: that it is the ego that is the true *subject* of the unconscious. In other words, that the apparently thinking and acting individuality of the ego (cogito) is nothing more or less than a subjection to some *thing*, to a thing that acts and thinks *in advance of it*. This is Žižek's obscene "secret" about Cartesianism,[4] and it returns in various guises throughout the last two decades in the question—as Lyotard frames it—of "*what if* what is 'proper' to humankind were to be inhabited by the *in*human?"[5] It is for similar reasons that I would venture to suggest that Humanism stands less in confrontation with the *in*human, than that it describes a *subjection* to the so-called *post-human*, and that (to rephrase Lyotard) Humanism is constitutionally and ceaselessly pregnant with its post-humanity.[6]

It is no accident that discourses of post-humanism *necessarily* engage with a certain humanistic tradition, whether it be called "enlightenment," "scientific rationalism" or—seemingly paradoxically—"technologism," and what we might call the "method" of knowledge, certainty, truth; in short, the very *technē* of human understanding. In this view, the human is regarded not as the instigator of particular technologies, but as a *prosthesis* of technology. The idea of man does not give rise to the deadly efficiency of the mechanised world, but is rather the prosthesis of that world—and here we must allow the word "prosthesis" to resonate with several apparently contradictory meanings, centred around the notion of extensibility and addition, as well as of something "placed before." The category of the human may be given as something more than merely a surplus produced by a system of technologisation bound to political economy, for example, and yet it shares the evanescence of the commodity and of spectacle, as a type of dream flickering across a "paralytic," entropic surface of mechanical reason.

If Humanism is in part that discourse which finds its expression most forcefully in the project of civilisation and technological modernity, then it is necessary to examine the implications of Žižek's formula against the accepted understanding of Freud's proposition that "civilization is based on the permanent subjugation of the human drives."[7] What would it mean, in other words, to treat the human drives as precisely this "excess in advent" (as Lyotard says)[8] of a generalised technicity—not what is subjugated by technology, for example, but rather what is expropriated to it? This is what it would mean to think "the human" as a *prosthesis* of technology.

And yet, of course, this thinking is not in any way alien to Freud's project, nor is it a novelty to introduce it to Joyce's. Luke Thurston, in his book *James Joyce and the Problem of Psychoanalysis*, provides one context for approaching the question of the post-human in Joyce, which can usefully be extended back through Donald Theall and Marshall McLuhan's investigations into Joyce's techno-poetics. Thurston points to a Dantesque technicity at work in Joyce's writing, redolent of a certain (Freudian) repetition compulsion, from which the concept of "will" emerges as a type of cybernetic circuit, tied to the Nietzschean conception of eternal return—or what Pierre Klossowski terms "the vicious circle principle."

It is within this genealogy that Thurston positions an as yet unnamed "post-humanism" in Joyce as the consequence—so to speak—of Nietzsche's "metamorphosis of the individual"[9]: "not as the transformation of individual human beings, but as the break with the very concept of self-identity–the *Principium Individuationis*–that Nietzsche borrowed from Schopenhauer to designate modernity's mechanistic expropriation and alienation from the primal chaos of nature."[10] Consequently, we witness a shift from the idea of technological modernity as expropriation and alienation of human identity and so on, to a technicity *of* self-identity. In other words, to the invention of the human *as* a movement of the post-human and of the "factification," as Joyce says in the *Wake*, of the Schopenhauerian *world-as-will-and-idea*.

Textual Neurasthenia

In a well-known analysis of the opening passage of Joyce's "Sisters," Hélène Cixous draws attention to precisely such a "factification" in Joyce by way of the operations of *paralysis*—which may be thought as naming a certain *neurasthenia of the text*.[11] This term, "paralysis," which haunts *Dubliners* and much of the critical literature surrounding it, implies a fantasmatic doubling of the analytical scene and of its hermeneutic—towards the interior and radically alien "identity" of this text, its "sinful and maleficent being," into which the so-called subject is expropriated by a series of ellipses, repetitions and displacements. "There was no hope for him this time: it was the third stroke." And so on. All of this is well known.

What is of especial interest here, however, is the propositional and causal structure that underlies this expropriation and points towards something like a rationale, even an operation approximating *reason*, in what at first appears to be nothing more than a paralytic technicity. "There was no hope for him this time: it was the third stroke." Throughout, the text supplies reasons. The reason that there was *no hope for him this time* is because it was the third stroke. What these terms mean is far from clear and yet, on balance, there appears to be no real uncertainty: everything that needs to be expressed is expressed through that colon. Reason is given: a post-effect. Not because it fills a gap, but because it *presupposes itself*, as it were, in the figure of the missing object; the occluded; the other. Precisely "it"—this thing that somehow thinks, in language.

> There was no hope for him this time: it was the third stroke. Night after night I had passed the house (it was vacation time) and studied the lighted square of window: and night after night I had found it lighted in the same way, faintly and evenly. If he was dead, I thought, I would see the reflection of candles on the darkened blind, for I knew that two candles must be set at the head of a corpse. He had often said to me: 'I am not long for this world,' and I had thought his words idle. Now I knew they were true. Every night as I gazed up at the window I said softly to myself the word paralysis. It had always sounded strangely in my ears, like the word gnomon in the Euclid and the word simony in the Catechism. But now it sounded to me like the name of some maleficent and sinful being. It filled me with fear, and yet I longed to be nearer to it and to look upon its deadly work.[12]

The word "paralysis" links the so-called subject to something like an *agency in language*, which is not a *signified agency* but rather the agency of signification "itself" and which appears to go about its business autonomously—a *telepathy* of "things" (the *materiality* of language) which seem miraculously to transmute a certain sense, or even nonsense, of the paradox at the heart of the approach to the Real (the infantilism of a Roquentin; the fascination of "paralysis" by other means).[13] This *thingness* suggests a type of golem of inscription (the "shem"), calling to mind the obscene, alchemical body of Joyce's doppelganger Shem towards the end of Book 1 of the *Wake*. Here, Joyce-Shem is depicted as producing:

> nichthemerically from his unheavenly body a no uncertain quantity of obscene matter not protected by copriright in the United States of Ourania or bedeed and bedood and bedang and bedung to him, with his double dye, brought to blood heat, gallic acid on iron ore, through the bowels of his misery, flashly, nastily, appropriately, this Esuan Menschavik and the first till last alchemist wrote over every square inch of the only foolscap available, his own body, till by its corrosive sublimation one continuous present tense integumented slowly unfolded in all marryvoising moodmoulded cyclewheeling history.[14]

What we encounter here is a "paralysis" which, as in "The Sisters," is no longer an inertia, an *immobility* or rigidification as such, but rather a perversion of the body as something *bound*, of the fixed materiality of the "text," in which the subject (which, in any case, is never a discrete entity) is confronted with a radical insufficiency: the body, the text, wills itself—so to speak—and henceforth the subject is seen to exist as a type of putrefactive superfluity. It is no longer that vehicle, that prosthesis of reason, but rather the fully fledged autonomous *thing*, the *undead*, the Freudian *partial object* metamorphosed here into a series of metonymy and synecdoche. Like the tongue, *playing dead* (as Cixous says), in the mouth of the priest-father who returns, in nightmares, in order to communicate something unsayable, obscene. The grey, disembodied face of the paralytic is also an ("undead") *thing*; an "it"—"*it* murmured"; "*it* desired to confess"; "I found *it* waiting for me"; "*it* began to confess"; "*it* smiled continually"; "*it* had died of paralysis."[15]

> In the dark of my room I imagined that I saw again the heavy grey face of the paralytic. I drew the blankets over my head and tried to think of Christmas. But the grey face still followed me. It murmured; and I understood that it desired to confess something. I felt my soul receding into some pleasant and vicious region; and there again I found it waiting for

> me. It began to confess to me in a murmuring voice and I wondered why it smiled continually and why the lips were so moist with spittle. But then I realised that it had died of paralysis and I felt that I too was smiling feebly, as if to absolve the simoniac of his sin.[16]

At every point, the "it" provokes and also determines the actions of the protagonist. At every point, the protagonist is in this way *situated*, even as the coherence of any subject remains in question, as though on the verge of a constant *re-covery*. As in the Freudian algorithm, where *it* was, there ego shall be. And there, too, will be that evanescent Humanism which is, as Sartre says, "the predicament of having to choose without appeal to any concept of human nature that would guarantee the rightness of his choice and the efficacy of his actions."[17] But more: because while it appears that the protagonist in this drama of human action is constantly putting words in the mouth of an absent "human nature" (the boy in Joyce's text is constantly playing at the game of [an impossible] *analysis* and thereby recovering the *posthumous* speech, the "desire" of the paralytic—i.e., to confess [its sin]); it is rather a question of this "human nature" *as that thing that (posthumously) returns* and which, in returning, *decides in advance of us*—the primal, originary character, we might say, of this "sinful being."

How can this be? Precisely because it is the perversely autonomous object—this suppressed or excluded "nature"—that gives the form of the human dilemma in the first place. The *terms* of the dilemma cease to be of primary concern; rather we are—as Sartre says—in the domain of a predicament "of having to choose and act"; it is not, then, a question of the rightness of the choice, but that, in being situated *as* the apparent necessity of action, the "human" describes a prosthesis of reason masked by indeterminacy. In a schema of generalised paralysis—in which the mark of the paralytic anticipates the subject at every point—a moment of indeterminacy appears to arise; an ellipsis is given to be filled; a parenthesis in which the subject is seemingly enabled to inscribe itself in the reality of "its own" actions.

This, of course, is the mark of the phantasmatic; the very contrary of Sartre's insistence that "there is reality only in action."[18] Its counterpart is that conception spelled out by Heidegger in his "Letter on

Humanism" (1947) where—among various other things—Heidegger argues that "Thinking acts insofar as it thinks. Such action is presumably the simplest and at the same time the highest, because it concerns the *relation* of Being to man."[19] Again, the human resolves itself into a type of object of a "thinking" which "acts," and whose acts do not belong to man, but instead concern "the *relation* of Being to man." It is an already *posthumous relation*; of what stands outside "it," as a *prosthesis of Being.*

Necessarily, this is a simplification. Nevertheless, let us return for a moment to Joyce's text. From the very outset, we are given a series of propositions in which a kind of "reason" is generalised by way of a "figure" of paralysis, between the "I" of the protagonist and the "it." It, we learn, "was the third stroke"; "was vacation time"; "sounded strangely in my ears"; "filled me with fear"—"and yet *I* longed to be nearer to *it* and look upon *its* deadly work"… Above all, "*it* sounded to me like the name of some maleficent and sinful being"… We notice, immediately, how the word paralysis operates in an environment in which all else appears suspended: time as repetition, as vacancy ("it was vacation time"); idleness ("I had thought his words idle"); deduction ("I knew that two candles *must* be set at the head of a corpse"); speculation ("I […] studied the lighted square of window"), and; revelation ("now I knew they were *true*"), reduced to geometric principles ("the word gnomon in the Euclid") and the relation of objects outside thought, etc.[20]

Finally, we notice that the word "paralysis" *operates.* Its environment is one in which signification constantly feeds back, recursively, metonymically, in the figure of the "it" attributed with all sorts of verbal features and whose generalised *materiality* is foregrounded. The word paralysis, the way it sounds on the protagonist's lips, the way it names some "maleficent and sinful being" that—despite the apparent signification of the term—*works,* links the nascent subject into an agency within language. And it is this agency—even if this may seem a contradiction—by which subjecthood is effected as *a relation of terms.* For it is solely in this relation that the *objecthood* of language is insisted upon; an objecthood in which meaning, knowing, thinking and being are conflated in a series of oppositions (for which reason Heidegger notes that "the essence of Humanism is metaphysical").[21]

Here, too, we see at work the logic of a Humanism which is the installation of a negativised *technē*—which Heidegger defines as "a process of reflection in service to doing and making."[22] A product, in other words, of a type of scientism and a "technical interpretation of thinking" (i.e. utilitarianism). The insufficiency of this definition of *technē* obtains from certain other insufficiencies in Heidegger's argument, most noticeably to do with the critique of the *animale rationale*, which Lacan and others have examined in some depth.[23] What is important here is to recognise in *technē* a structural operation—of *relations*, as Heidegger says—by which any Humanism is necessarily underwritten, even if this is solely in the form of a dilemma. (It is for this reason that we cannot simply limit post-humanism to a type of transcendental cybernetics—the sublation of human/machine or human/animal—it is not a matter of transcended dichotomies or of a gesture of negation.)

And here, too, resides the apparent paradox of my thesis, according to which agency is effected (or rather affected) in the figure of the Cartesian subject as a type of deadly work—the work of paralysis "itself," of language "itself," of that thing which, as Lacan says, constantly returning (as in dreams), *thinks*.[24] The human obtains in the predicament of having to choose only through a mode of deferral—a mode which is the composite of a desire for the obscene in which it, the "subject," is both *posited* and *rationalised*. This reflective movement, in which Humanism emerges as a post-effect of a generalised (and not simply utilitarian) technicity, recalls again the very familiar scenario underlying the protagonist's dream of the paralytic in Joyce's text. We recognise a human eye turned upon a kind of *afterlife*, where consciousness seems to live on after death, to be omnipresent, to speak from a kind of abyss, so that through "it" we may bear witness—paradoxically—to our absence, as the (dis)embodiment of a kind of *thinking* that may be said to bear witness to the end of thought, to the unthought.

Notes

1. Jean-Fançois Lyotard, *The Inhuman: Reflections on Time*, trans. Geoffrey Bennington and Rachel Bowlby (Stanford: Stanford University Press, 1992), 1.
2. Jacques Lacan, *Le Séminaire XXIII: Le sinthome* (Paris: Seuil, 2005).
3. Lacan, "Du sens, du sexe et du réel," *Le Séminaire XXIII*, 119.
4. Slavoj Žižek, *Cogito and the Unconscious* (Durham: Duke University Press, 1998).
5. Lyotard, 2.
6. Lyotard, writing of "modernity," is especially attentive to the "pointlessness of any periodisation of cultural history in terms of 'pre-' and 'post-,' before and after, for the single reason that it leaves unquestioned the position of the 'now,' of the present from which one is supposed to be able to achieve a legitimate perspective on a chronological succession" (24). It is important to add that the tropological character of the post, as in post-humanism, is necessarily in the order of a "metonymic forethrow" which in fact underwrites any conception of a "now".
7. Herbert Marcuse, *Eros and Civilisation* (Boston: Beacon, 1955), 3.
8. Lyotard, 25.
9. Pierre Klossowski, *Nietzsche and the Vicious Circle*, trans. Daniel Smith (Chicago: University of Chicago Press, 1998), 71.
10. Luke Thurston, *James Joyce and the Problem of Psychoanalysis* (Cambridge: Cambridge University Press, 2010), 5.
11. Hélène Cixous, "Joyce: The (R)use of Writing," in *Post-structuralist Joyce: Essays from the French*, eds. Derek Attridge and Daniel Ferrer, trans. Judith Still (Cambridge: Cambridge University Press, 1984), 23.
12. James Joyce, "The Sisters," in *Dubliners*, ed. Robert Scholes with Richard Ellmann (New York: Viking Press, 1967), 1.
13. Cf. Lacan's "télépathie émetteur" in "Joyce et les paroles imposes," *Le Séminaire XXIII*, 96.
14. James Joyce, *Finnegans Wake* (London: Penguin, 1992), 185.29–186.02.
15. Joyce, "The Sisters," 3.
16. Ibid.
17. As per Martin Heidegger, "Letter on Humanism," in *Basic Writings*, ed. David Farrell Krell, rev. ed. (London: Routledge, 1993), 214.

18. Jean-Paul Sartre, "Existentialism is a Humanism" (1946), in *Existentialism from Dostoyevsky to Sartre*, ed. Walter Kaufman (New York: Meridian, 1989), 62.
19. Heidegger, 217.
20. Joyce, "The Sisters," 1.
21. Heidegger, 247; Heidegger further adds: "In this regard 'subject' and 'object' are inappropriate terms of metaphysics, which very early in the form of Occidental 'logic' and 'grammar' seized control of the interpretation of language" (218).
22. Heidegger, 218.
23. See, e.g., Jacques Derrida, "And Say the Animal Responded," in *Zoontologies: The Question of the Animal*, ed. Cary Wolfe, trans. David Wills (Minneapolis: University of Minnesota Press, 2003); Heidegger: "Expelled from the truth of Being, man everywhere circles around himself as the animal rationale" (245).
24. Lacan, "La chose freudienne, *ou* Sens du retour à Freud en psychanalyse," *Écrits* (Paris: Seuil, 1966).

3

From Law to Code: Posthumanism as Sinthome

Judith Roof

Does humanism's focus on ethics, rationality, personal rights and liberties, creativity, and individual insight depend upon the metaphorical symbolic structure (Law), the sinthome, the topological version of the symptom that links the real, the symbolic, and the imaginary? Does a shift in the character of this sinthome from a reliance on speech and metaphor to a belief in the metonymic primacy of matter (an imaginary misunderstanding of the real) account for the emergence of object-oriented and material ways of thinking that reduce the human subject to object? The "symptom" is a "metaphor," a part of language, a signifier.[1] In *Seminar XXIII* (1975), Lacan locates the symptom as the subject's mistake, a stumbling block that signifies a search for meaning in speech and whose formation eventually links the three orders—the symbolic, imaginary, and real—as a way to subtend the subject.

J. Roof (✉)
Rice University, Houston, TX, USA
e-mail: roof@rice.edu

S. Matviyenko and J. Roof (eds.), *Lacan and the Posthuman*, The Palgrave Lacan Series,
https://doi.org/10.1007/978-3-319-76327-9_3

The "symptom" that amends the subject also serves as the "sinthome," a fourth ring of the topological Borromean Knot that binds the other three rings together in the mischances of meaning. This fourth ring has long been, like the symptom itself, linguistic and metaphorical. The Law, as sinthome, works through metaphor, as the Law-of-the-Name-of-the-Father, the seal the Oedipal configuration of the three Borromean rings. Recently, (i.e., in the past century) the guarantor of Oedipal limit and relation has shifted increasingly to the molecular and material; the sinthome as the fourth binding ring operates increasingly metonymically in the guise of the molecular "matter" upon which contemporary theories of ontology rest, providing a new version of "the enigmatic link between the imaginary, the symbolic and the real" that offers the "myth" that sustains subjectivity "as real."[2] As Lacan maps in *Seminar XXIII*:

> To speak as Chomsky does, see if you will what this genetic real is all about. Language only enables it to be approached in terms of signs, in other words, messages. The molecular gene is reduced to what secured the reputation of Crick and Watson, namely the double helix from which the various tiers that organize the body at a number of different levels are deemed to set off. First there is division, development, and cellular specialization. next there is specialization based on hormones, which act as vehicles for all sorts of messages that direct organic information. With all these so-called messages, we arrive at a great spiriting away of what the real actually involves.[3]

In the contemporary moment, the metaphor that undergirds Law and supports Humanism's ideals—the sinthome of the Name-of-the-Father—is increasing becoming a sinthome reliant on a metonomy itself premised on the imaginary real of "matter." This shift displaces the human into an imaginary bifurcation in which the subject objectifies itself as a part of a larger material network and simultaneously occupies the site of the all-seeing subject: an imaginary scalar perspective from which the now-veiled human subject can envision and define such a network from the dual perspectives of molecule and God.

Untying the Knot: Metaphor as Sinthome

> I'm saying that what forms the Borromean link has to be supposed to be tetradic—that perversion merely means *version vers le père*, a version towards the father—and that all in all, the father is a symptom, or a sinthome, as you wish.[4]
>
> Jacques Lacan, *Seminar XXIII* "The Sinthome"

Where do we begin to outline this long shift from the sinthome as a linguistic brace for the Oedipal three of the Borromean knot to a sinthome seemingly comprised of the metonymical imaginary of "real" matter? Let us begin with the Law. In the thirteenth century, two hundred years after Anglo-Norman William, the Conqueror invaded England and wrested rule from its Anglo-Saxon incumbents, the English system of "Common Law" was well in place in the territories ruled by the Plantagenets. Unlike the Civil law of the Romans, which operated by means of a code of laws enacted by sovereign ruling bodies, English common law was the accumulated wisdom of generations of court decisions. These "opinions" then governed ensuing cases (disputes brought before a court) under the principle of *stare decisis*, which strongly encouraged judges to follow the decisions made by previous judges. Despite its deep history in Anglo-Saxon culture, English common law was formalized by the Plantagenets, particularly Henri II, though the very last piece, the Magna Carta, which declared that the monarch had to follow the same legal principles as everyone else, was enacted during the reign of Henry's son, John.[5]

English common law offers three exemplary stagings of Lacan's of the Borromean enknottedness of the symbolic, imaginary, and the real, fronted by the sinthome offered by the metaphor of the "Law-of-the-Name-of-the-Father": (1) The "title" to property, or the "Deed." (2) "*Voir dire*," the name still in use for a jury procedure, denominated by its guaranteeing admonishment, which enacts the very structure of the relation between language and the real. And (3) "*Stare decisis*" as the performative coincidence of the imaginary and the symbolic in language that defines the fate, status, acknowledgment of a real it ever edges away from. Metaphor in the guise of the Name-of-the-Father reknots the lacuna between language and the real that subtends the Law's symbolic power.

A word that makes one of its first appearances in English common law is the word "real," from the Anglo-Norman "*reel*," with an etymology that goes back to the Latin "*res*" or "thing." In English, the word "*reel*" was first applied to "property" as in "real property," a legal category that refers to a circumscribed chunk of land as opposed to, say, jewelry, cattle, or stocks and bonds. Appending the adjective "real" to property doubles what we already think we know about property—that it is real. The modifier "real," however, is necessary because the word "property" does not refer to the land at all, but to something else—to the ownability of the thing. The word "property" comes from the old French *propriété*, which means proper to, belonging to, an aspect of. Using the term "property" in the place of any direct signifier of the terran "*res*" thus signifies the ownability of the *res* instead of the *res* itself. But this overcompensatory motive may also come from another fact: The relation between an owner and real property is always imaginary, secured by the symbolic appendage of a name. This suggests that real property itself can become an *objet a* in a dialectic of desire.

There is no intrinsic relation between a human being and a piece of land. And the "real property" itself, which is a piece with specific arbitrary boundaries, is an imaginary construction. Thus, the formation "real property" already requires two imaginary leaps: First, the imaginary lines that demarcate whatever stretch of land constitutes real property; and second, an imaginary relationship between a human being and this imaginary bit of terrain. "Real," then, which is really a version of "reality" as opposed to the "real" of the Borromean knot, modifies this doubly imaginary relation, which, in the end, is both ambiguous and doth protest too much. Denominating any "res" as real makes it really real (as well as correspondingly less real) as it locates the "res" soundly in the arena of "sens," the site of *jouissance* in the overlap between the imaginary and the symbolic.

But for real property to be legally propriété, a third process is necessary: a deed. A deed, *l'acte notoraire*, is a document that secures the relation between a human being and a piece of land as a matter of law. This deed document, derived from whatever symbolic power holds sway, offers a symbolic stamp upon imaginary relation, operating as a Seal, as a guarantee in the form of a small image (yet again, old French). This Deed sutures a Name—the owner's—to a piece of land with the accord of the sovereign signified by an image, name upon name.

Like the Deed, the name of the father is a guarantee that certifies the reality of an imaginary biological relationship while signaling its contingent character. So the real, misunderstood as material, whether that is tangible immovable land or a child, always exists as a part of a relation—name appended to land or child, and never exists in itself in any form to which we have access outside of language. Millennia of paternal anxieties about irrelevance and parasitism produce and model a knotted Law which requires those three terms: an imaginary real in the form of land or flesh or perhaps any tangible "res" (and whether the real has to be tangible is yet another question), an imaginary proscription that takes the form of boundaries or authorship or causality, and a symbolic guarantor—a writing or Name—that redoubles the imaginary relation between a human subject and the real, whatever that real is. The Law-of-the-Name-of-the-Father, thus, is the quintessence of metaphor, where a signifier stands in the breach between signifier and signified, a subject and the imaginary realities of causality and foundation, producing and certifying a relation where there may be none.

This metaphor is the principle of almost all language in the Saussurian arbitrariness of the link between signifier and signified as well as the Ur-Law that grounds Law as enactive metaphor: Lacan's "Law-of-the-Name-of-the-Father." Since we never knew for certain who the father of a child was, the Name produced the relation in the same way as the deed produces ownership (or vice versa). The law of the Law, metaphor, defines the operation of Law itself as the sinthome, the fourth "ring" that appears to retie the Borromean knot. This sinthome veils the "hole" that exists between the real and the symbolic/imaginary of ownership in an operation that plays out almost too literally what the fourth ring sinthome accomplishes in securing the Borromean knot. Had the sinthome not been metaphorical speech in the form of the Name, it might already have been metonymy in the guise of matter—or *mater*—waiting in the wings with a metonymical relationship to the infant.

The discovery of DNA and the mapping of the Human Genome should not, theoretically, alter the operation of the metaphorical Name as a sinthome at all, since even now DNA provides only a statistical likelihood of paternity. Until the late 1980s, there was never much certainty about who Daddy was, but from the 1980s on, DNA testing enabled the discernment "to a high degree of certainty" who the father of a specific child is, assuming there candidates in mind in the first place.[6] We now pretend to treat DNA the way we previously treated the Name-of-the-Father—as a material guarantor or sinthome retying the Borromean knot. It is worth noting that in order to sell the Human Genome Project to the public—and in order to disseminate much of anything at all about DNA—it was necessary to metaphorize the genome via the analogy of letters, words, and histories—just like the father. So if we were to append a genetic origin to a child, it might be the same as appending a name, in so far as the genome itself has already been metaphorized as a language. Nom du Père, Nom du DNA. Nothing has changed in relation to the real except the sinthome which seems to have shifted from the Sovereign to Science, from metaphor to metonymy, imagined to exist in the "materiality" of DNA, which though it might seem to approach the real, is no more real than a name.[7]

Though we tend toward doing so, we do not yet append genomes like names. The sinthome of material metonomy has not quite taken over even though we no longer need the Name of the Father to produce relations. When it is necessary to know who the father of a child is, we conduct tests of chromosomal fragments that give us a likelihood—a percentage probability and by no means a certainty—that a given individual is the father of a particular child. So the name, which no longer necessarily produces a relation where none provably exists, nonetheless still—and anachronistically—hovers as a ghostly sinthome, allied with a second, science, the discourse of metonymies, of matter's relation to matter, that Lacan compared to a "Hysteric's discourse." The two sinthomes offer a double seal when things are in doubt.[8] A double knot retying the Borromean triad. Things have gone cubic, 3-D. The Law-of-the-Name-of-the-Father was the sinthome that addressed the origins of the subject; the sinthome of scientific matter turns the real into Truth.

Voir Dire, or Compensatory Posthumanism

> ... there are even people, science fanatics, who tell me Keep on trying to know. What's this about? Of course you must say what you know about the names of the father.
>
> Jacques Lacan, *Seminar XVII*, 143[9]

Every time we try to evoke the real, it slides metonymically sideways, leaving a trail of associated terms. The version of sinthome as the Law of the Father appeared to connect the symbolic, imaginary, and real via what Lacan calls "quilting points or *points du capiton.*"[10] The emerging formation of sinthome of materiality and metonymy is much more worried about the verity of the matter with which it fronts itself, confusing matter with reality. Getting to this reality, however, as Lacan's Borromean logic suggests, is like trying to grasp a bubble or see the back of your own head. And no matter what scale we occupy, the reality of matter is always the wrong scale—too large like the universe or too small like a sub-atomic particle, if even we understand this matter as the same matter as that comprising the sinthome. The extremity of the scale problem does not mean that matter qua reality is somehow foundational (exists at either end of what we can comprehend as existence), but that it is, like the real, non-representable in itself. As the real slips metonymically through representations, we still try to evoke the real in the mistaken form of "material" as the stable point of a perpetuated metonymy that we confuse—again via metonymy—with the truth, an imaginary synonym, yet not, for both matter and the real. What, finally, is science, but a quest for the real as Law, misunderstood as tangible, true, foundational, and originary? The slippage in the character of the linking sinthome from metaphor to metonymy represents a delusional shift from language to "truth" as that which ties orders together.

Just as the Deed deployed the name to produce a relation where none necessarily exists, so the legal oath guarantees truth. Like the name, an oath is a linguistic guarantee, but in so far as oaths bear on matters of matter, they illustrate the ways "matters" before the court become questions of truth about the "matter" that proves such truth. In most causes of action in the American legal system, defendants have a right to a trial by a jury of their peers. The word "jury" comes from the old French *juree* meaning "oath." "Oath" comes from Old English, meaning a "solemn appeal to deity in witness of truth or a promise." In its appeal to language that operates metaphorically as a guarantee, the oath is analogous to the deed. Those impaneled on a jury in the USA undergo a process called "voir dire" in which the lawyers for each side question jurors to determine which ones might be unusually biased or unreliable. The old French phrase "voir dire" stymies American courts, who in attempting to explain the process, usually translate the phrase "voir dire" as "to see, to say."[11] This translation is completely incorrect, but in a nicely symptomatic way. It appeals to the delusional belief in eyewitness credibility, to a trust in an unassailable correlation between experience and representation, as if the translation between the two offers no deviation—no metonymy—from the reality of experience at all. Juries can say what they see with nary a fissure nor a slide sideways. They guarantee that their language will not be metaphorical, both in its relation to matter and as figures of speech.

But "voir dire" is actually even more symptomatic; the phrase really translates as "to speak truly," or "to tell the truth." *Voir* is a form of the old French "*veir*," meaning "true" from the Latin *veritas* and this word is the next metonymic stop on the real's slide toward matter (and vice versa). The real slips into the Truth, which is increasingly based upon observations about matter. Obviously, there is a shift of meaning here; "real" and "Truth" are not synonyms. What is true may be different from the real, the Truth doesn't have to be real, but we rarely interrogate the slippage in that metonymy. Juries look for the truth, whose relation to the cause at hand is one of discerning among arguments and the positioning of "material evidence." If they survive voir dire, jurors generally take an oath that they will deliver a verdict according to the evidence. And here we are again. "Verdict" means exactly the same thing as "voir dire." As with "Real property," the insistence of certain terms—here "*veir*" or "*voir*" is suspiciously compensatory. Jurors who speak truly deliver a true verdict based on the evidence, which is a much if-ier word that has its roots in the Latin *evidens* which means "obvious" or "apparent," a term for appearances.

Stare Decisis, Or Matter Matters

> Indeed, interpretation operates solely through equivoques. There has to be something in the signifier that resonates.
>
> It comes as a surprise that this wasn't even remotely apparent to the British philosophers. I'm calling them philosophers because they are not psychoanalysts. They staunchly believe that speech doesn't have any effects. They're wrong.
>
> Jacques Lacan, *Seminar XXIII*, 9[12]

Stare Decisis—that courts should follow the precedent set by previous courts—embodies the ways the language of written court decisions constitutes itself as law—as the symbolic that governs how courts decide other similar matters. It is a perfect example of the Law-of-the-Name-of-the-Father as the sinthome that distributes the language of court decisions, describing, treating, recounting, resolving one circumstance which presumably occurred in "real" time among "real" people—i.e., as a sort of imaginary replay of the problem it resolves. The principle from one case becomes the rule for similar situations that follow.

Stare Decisis sets formulae, not unlike mathematical formulae, but in language instead of symbols. Derived from what we might regard as an experiment of sorts—humans in a tussle over whose principles define the evidence (which in contemporary life becomes increasingly molecular)—*Stare decisis* is, like geometry, an early form of a scientific method in which the principle of the arrangement of principles both describes and defines another principle, which then, ad infinitum, defines other principles and the way we understand the relations among principles as a form of precedence and Law as a process of discerning analogies.

Courts describe causes of action as "matters before the court," the evocation of "matter," a telling hint to lurking sinthome of matter understood as tangibility, objecthood, objectivity, the "res." And in Law, *res ipse loquitur*.

But matter in the form of "trace" or "forensic" evidence of all kinds—DNA, hair and fibers, fingerprints, saliva, blood-grouping and serum-protein and enzyme typing—have begun increasingly to matter, depending on the probative value of their presence, the reliability of the protocols for gathering, typing, and comparing matter, and the degree of probability enabled by material "matches."[13] Such "physical" evidence, at least in the minds of general publics now accustomed to the plethora of matter-evidence television shows such as any *CSI*, competes with language, testimony, and other evidential modes as often *the* determining factor of a Truth now understood as molecularly evident. Outstripping testimony, legal principle, and argument, the material thing indeed has come to speak for itself. It is becoming the sinthome. Science, imagined as a reality, trumps other modes of possibility.

But just as courts of law make matter matter, physics begins to see that matter is not necessarily material. Since Einstein, matter qua matter has become something else, energy, so that whatever we might have imagined was a material basis of something like the real is not itself real at all in the sense that this reality remains tangible and objective. The imaginary reality of matter retreats along with scale. The closer we look, the less definitive matter is, from molecules to atoms to atomic particles to quarks, bosons, and leptons, to techni-quarks to … ? Whenever physicists think they have gotten to the smallest scale bottom of things or largest scale outside, there is yet another scale. The reality of matter recedes infinitely both directions, if we even think of the real as making up matter (or matter as comprising reality) in the first place. If legal systems depend on science to discern reality (or as we have seen above, Truth) then it has looked the wrong direction, though to be fair, physicists look for the theory—for the symbolic/imaginary formulae that might explain the behavior of the real and not the real itself. Reality is not in the matter. Our descriptions and belief in matter are a way to stave off the real by thinking we have defined it.

Objects Relations

> There is no hope of breaking, in any way whatsoever, the constitutive knot of the symbolic, the imaginary, and the real. In this respect, let's put it in the most coherent way we can, it passes up what constitutes a virtue, a virtue that is even said to be theological. This is why our analytic appreciation of what is involved in the knot is the negative of religion.
>
> We no longer believe in the object as such. It is in this respect that I reject that the object can be grasped by any organ.
>
> Jacques Lacan, *Seminar XXIII*, 25–26[14]

With the shift in the character of the sinthome from language to the imaginary of the material, the role of language has changed from the capacity to produce relations between unrelated things to a merely descriptive and infinitely translatable-to-image vector of objectifying commercial coordinates. Having become transparent—we mean what we say and we can say what we mean—language becomes a mere tool in the realm of the material. With this shift, we forget that language is always more than it says and that consciousness of being itself is available only through metaphor (language, image, gesture). Premised on metaphor, humanism begins to give way as the material imaginary of metonymical systems—digital coding, molecular activity, DNA—appears to offer a more definitive understanding of the relations among objects. In this process, individual subjects become objects as well, no better or worse than other objects occupying the same space.

But this shift in sinthome has not happened suddenly and is still not completely over. Hints of desperate attempts to revivify the Law-of-the-Name-of-the-Father in ironically literal terms emerge, all of which try to elevate the fiction of phallic supremacy (the fiction that required the Law-of-the-Name-of-the-Father in the first place). The loss of "Law" inspires a gang of compensatory and overly literal phallic gestures (the wide distribution of firearms, a proliferation of pickup trucks, restrictions on birth control and abortions, bellicosity, the broad and occasionally pugnacious re-emergence of spectacularly patriarchal religions) to moor up and pretend the metaphorical relation between name and connection that seemed to privilege the male still operates.[15] Signs of a desperation to re-erect the Father's Law, understood as privilege mistakenly thought to emanate from the Phallus qua penis instead of representing exactly what the Phallus, as signifier, is not, these attempts to valorize the literal character of an old order take male privilege as a literal truth instead of as the compensatory process it always was in a world where no one knew for sure who the father was. The Law grounded in the metaphor that made a connection where none existed in the real becomes the law based upon the mistaken notion that the symbolic was ever "truth" in the first place.

These clumsy cultural compensations coexist with the shift to metonymical binary code systems (digital machines, algorithms) that offer the illusion of a focus on the user's objectified "self" as the intersection of countless connections organized via the metonymical sinthome's new "law" of scattered relations among all things. Sustained by algorithmic platforms designed to generate images within an imagistic network, this new sinthome's feigned imaginary covers over the "foundational hole" with the ersatz provision of mimo-graphs disconnected by definition from the real and the symbolic they progressively displace. This "code" system offers narcissistic commodity logics in place of performative language, encouraging a perpetual narcissistic timeout trapped in the imaginary. The Humanist subject becomes the metonym's object in a system whose only "law" is the illusion of user gain in a mechanism of covert plunder. Users (and now we all have little choice but to give into machines in their ersatz guise as "tools") become what the metonymical machines feedback loops tell us we are. The machine focuses desire on the self whose status as a "self" is produced as an imago instead of as a complex relation with the "Other." The "Other" has become a set of "helpful" algorithms and hence, in a narcissistic short circuit no longer "Other" at all. The old "Law" by which the name produced a connection between the real and a subject has become the lawless objectification of subjects who lack (and perhaps even consciously eschew) all of Humanism's attributes in pursuit of an imaginary order that characterizes itself as a distributed and egalitarian delusion of empowerment.

This sinthome of coded and reified matter has emerged in the past twenty years in academe as subtending "Posthumanism" as well as two additional "Posthuman" approaches to the loss of language as metaphor—the "New Materialism," and "Object-Oriented Ontology."[16] These latter constructions premise their questions and analyses on the belief that matter matters and that we have some unimpeded immediate access to that matter.[17] Both try to supplant Humanism's ideals in favor of a flat egalitarianism beyond the human that provides in itself a performative ethics. The New Materialism, which tries to avoid "the linguistic turn," is "[d]efined around the primacy of matter and its properties and actions, … it re-works long-held assumptions about the nature of the stuff of the universe. It responds to the need for novel accounts of agency, nature and social relationships in the contemporary epoch, when new questions have arisen about our place as embodied humans in the world and the ways we produce, reproduce and consume our material environment."[18] Both New Materialism and OOO seem to assume, as do many social scientists, that representation is a transparent process—that there is neither symbolic nor imaginary nor even human psychic processes at work in the apprehension of whatever this matter is. The unconscious disappears. Humans are merely Actors (or more accurately "objects") in a network among other causes, to appeal to Bruno Latour's version of this, automatons who unwittingly play their role and whose only critical stance is to observe and describe faithfully their place among the objects.[19] Going back to Lucretius, versions of the New Materialism objectify (a process we already know is only imaginary—who is doing the objectifying? Who can imagine themselves as not themselves except from the point-of-view of themselves?) around the primacy of matter which in itself seems to have agency, and which, as we have seen, can only ever be imaginary in the first place in so far as it comes, as Alfred Jarry demonstrated more than 100 years ago, from a reverse anthropomorphism—a turning matter into conscious agents.[20]

Object-oriented ontology also produces the imaginary of an existence in which objects and humans coexist as equal participants.[21] Both of these ambitious object-centered theories require the projection of imaginary consciousness, agency, or even abrupted subjectivity onto things and matter itself in a necessary and disavowed anthropomorphism. Both assume an equivalence between matter and the real as a way then to transpose the real into an imaginary that displaces the symbolic in a denegation of language which is the only means by which such notions could ever have appeared in the first place.

Posthumanism is a symptom—a stumbling point signifier whose attempt to merge the linguistic and the material signals the shift to a metonymical sinthome. What is "Posthumanism" but defense against and compensation for a pretend loss of the connection between language and being as well as the humble cover for an oddly hubristic enlargement of human capacity in the shift from perceptions imagined to be within the horizons of human subjects to a horizon that can look back upon the human as one phenomenon among many. In its various forms, Posthumanist thinking compensates conceptually for the loss of metaphor by erecting an even more extensive perspective in place of the human. Imagining the human as object, Posthumanist thought perceives the human from an imaginary point beyond the human, which in turn both produces and denies the illusion of subjective humility, while transforming the subject into the object of its own gaze, and ironically empowering its vision by denying its perspective as inescapably human. The sleight-of-hand that produces this instant egalitarianism operates by displacing an imaginary detached perspective away from the human subject so that the human subject can envision itself from a perspective imagined to constitute an improved, more masterful scale of vision, at the same time eliding the fact that the source of the perspective is a still specifically human—and a human whose perspective is now markedly extended. This imaginary shift in perspective results in a doubled mirror stage, the subject now standing between the reflecting mirrors of its two perspectives seeing itself into infinity. Posthumanism transplants Humanist values—ethics, rationality, insight, with an added dose of false humility—into the imaginary of this enlarged context, defined by systems that iterate the binary imaginary of digital universes, and that install a continued human aegis in the denial of its aegis—the obverse of the feedback loop that appears to center human users while they are being used.

Even the metonymical sinthome's elevation of machinic capabilities obscures the human hubris that undergirds the rules of algorithmic operation that produce simulations in a projected space that pretends to be disinterested. Algorithms, of course, are never "objective" or "inhuman," but always derive from sets of assumptions and biases that define their production by human programmers in the first place. Machines obey protocols designed by humans just as those same humans pretend humility in the face of the machine whose metonymical operations appear to extend beyond human capability—or in triumphant display, appear to display what contemporary humans conceive of as human capability. In bowing to metonymy, the magic of metaphor disappears into fantasy, the imaginary of a symbolic process elsewhere, enabling the illusion of a machinic universe of delusive presence to inhabit the emptiness of being. The metonymical sinthome wrests language itself from the realm of the human to the realm of the simulacrum, transposing the human from subject to subjectified, from originator to receptive object even as these two sinthomes—metonymical code and metaphorical Law-of-the-Name-of-the-Father—coexist, the human in the process of searching for whatever mode can offer closure as Law literalizes into "information."[22]

As the symptom of the metonymical sinthome, Posthumanist ways of thinking have not replaced or succeeded Humanism, but coexist with Humanist concepts in a rocky contest around the status of the subject itself. While the former eschews the Human in favor of a larger more systemic view, Humanism still scrabbles to define human rights in a world that tramples everything for profit. The lack of metaphorical language comprising Law makes it a free-for-all, sometimes subject to the sanctimonious rectitude of those who stalwartly believe in the binary oppositions such a loss of Law produces, unwittingly reproducing on a human scale the algorithmic operations of the machine.

Humanism struggles when the human subject becomes its own object and so do human beings. But why should we care? Doesn't Posthumanism show us why Humanism is no longer desirable? Alas, it shows us only the tensions attendant to competing sinthomes—metaphor (Law) and metonymy (code)—which produce competing notions of the subject—an enfranchised, conscious, linguistic subject and a disenfranchised, coded, object. The collapse of metaphor produces what is essentially a delusive one-way process of subjectification, in which the metonymic sinthome sustains a perpetuated automata, a retreat into the literal imaginary taken as somehow commensurate with being and meaning. But privileging the coded imaginary over signification at this point in time is itself a symptom of the very problem of the sinthome with which we now are faced. We now know, more or less, who Daddy is. Science displaces metaphor and the model of Law has been undone. Where to go? Ah, matter. It has been here all along. Science would tell us that all matter obeys rules without discrimination. If we are matter, then we, too, behave like matter. We, too, can be programmed, manipulated. Wishful thinking for some. Present in social thought at least since Marx. False analogy, alas.

Notes

1. Jacques Lacan, "The Agency of the Letter in the Unconscious," *Écrits: A Selection*, trans. Alan Sheridan (New York: W. W. Norton. 1977), 175. See also Valentin Nusinovici, "Symptom/Sinthome," *International Dictionary of Psychoanalysis*, encyclopedia.com, accessed August 24, 2017, http://www.encyclopedia.com/psychology/dictionaries-thesauruses-pictures-and-press-releases/symptomsinthome.
2. Jacques Lacan, *The Seminar of Jacques Lacan, The Sinthome, Book XXIII*, ed. Jacques-Alain Miller, trans. A. R. Price (Cambridge: Polity, 2016), 27.
3. Lacan, *Book XXIII*, 21–22.
4. Ibid., 11.
5. The history of the development of English common law is, of course, much more complex than this thumbnail. See, for example, Sir Frederick Pollock, *The History of English Law Before the Time of Edward I, vol. 1* [1898], accessed August 10, 2017, http://oll.libertyfund.org/titles/pollock-the-history-of-english-law-before-the-time-of-edward-i-vol-1.
6. DNA profiling techniques developed in the 1980s offered a better statistical chance of determining paternity, though no DNA testing is still 100% certain.
7. See Judith Roof, *The Poetics of DNA* (Minneapolis: University of Minneapolis Press, 2005). Recent studies of DNA such as Siddhartha Mukherjee, *The Gene: An Intimate History* (New York: Scribner, 2016), shift the metaphors deployed for DNA from language to "information."
8. In *Television*, Lacan declares: "I conclude that scientific discourse and the hysteric's discourse have almost the same structure, which explains our error, induced by Freud himself, in hoping that one day there would be a thermodynamic able to provide—within the future of science—the unconscious with its posthumous explanation."

Jacques Lacan, *Television*, trans. Dennis Hollier, Rosalind Krauss, and Annette Michaelson, ed. Joan Copjec (New York: Norton, 1990), 19.

9. Jacques Lacan, *The Seminar of Jacques Lacan Book XVII*, trans. Cormac Gallagher, accessed August 25, 2017, http://www.lacaninireland.com/web/wp-content/uploads/2010/06/THE-SEMINAR-OF-JACQUES-LACAN-XVII.pdf.
10. In "Subversion of the Subject and the Dialectic of Desire," Lacan defines this as the "anchoring point (*point du capiton*) by which the signifier stops the otherwise endless movement (*glissement*) of signification." (303) *Ecrits: A Selection*, trans. Alan Sheridan (New York: Norton, 1977), 292–325.
11. This was my own experience as a prospective juror in the Monroe County, Indiana courthouse. Not only the judge but the ABA video instructions for jurors translated the phrase "voir dire" as "to see, to say."
12. Lacan, *Seminar XXIII*, 9.
13. See "The Evaluation of Forensic Evidence," *NCBI*, accessed August 25, 2017, https://www.ncbi.nlm.nih.gov/books/NBK232607/.
14. Lacan, *Seminar XXIII*, 25–26.
15. The pickup market grew by 6% in 2016, and constitutes 15% of the total "light vehicle" market. Anton Wahlman, "2016 Statistical Analysis of the Pickup Market Wars: GM Vs. Ford Vs. FCA Vs. Toyota Vs. …" *Seeking Alpha*, January 15, 2017, accessed August 25, 2017, https://seekingalpha.com/article/4036949-2016-statistical-analysis-pickup-market-wars-gm-vs-ford-vs-fca-vs-toyota-vs.
16. Cary Wolfe carefully traces the origins of the term "posthumanism," also noting its different interpretations. As he formulates his own perspective, "Far from surpassing or rejecting the human—[posthumanism] actually enables us to describe the human and its characteristic modes of communication, interaction, meaning, social significations, and affective investments with *greater* specificity once we have removed meaning from the ontologically closed domain of consciousness, reason, reflection, and so on. It forces us to rethink our taken-for-granted modes of human experience, including normal perceptual modes and affective states of *Homo sapiens* itself, by recontextualizing them in terms of the entire sensorium of other living beings and their own autopoietic ways of 'bringing forth a world'—ways that are, since we ourselves are human *animals*, part of the evolutionary history and behavioral and psychological repertoire of the human itself." *What Is Posthumanism?* (Minneapolis: University of Minnesota Press, 2010), xxv.

In contrast N. Katherine Hayles sets out four attributes of posthumanism: "First, the posthuman view privileges informational pattern over material instantiation, so that embodiment in a biological substrate is seen as an accident of history rather than an inevitability of life. Second, the posthuman view considers consciousness, regarded as the seat of human identity in the Western tradition long before Descartes thought he was a mind thinking, as an epiphenomenon, as an evolutionary upstart trying to claim that it is the whole show when in actuality it is only a minor sideshow. Third, the posthuman view thinks of the body as the original prosthesis we all learn to manipulate, so that extending or replacing the body with other prostheses becomes a continuation of a process that began before we were born. Fourth, and most important, by these and other means, the posthuman view configures human being so that it can be seamlessly articulated with intelligent machines. In the posthuman, there are no essential differences or absolute demarcations between bodily existence and computer simulation, cybernetic mechanism and biological organism, robot teleology and human goals." *How We Became Posthuman: Virtual Bodies in Cybernetics, Literature and Informatics* (Chicago, University of Chicago Press, 1999), 2–3.

17. While the New Materialism locates "matter" as a kind of monistic first cause that defines the interactions of everything, OOO envisions a world in which everything of the Subject without any particular privilege. The problem with both is that the categories of matter and "objects" are human conceptions that do not even begin to address the wider range of phenomena that bring the nature of "matter" itself into question. See for example, Adam Frank, "Minding Matter," *Aeon*, accessed August 25, 2017, https://aeon.co/essays/materialism-alone-cannot-explain-the-riddle-of-consciousness, who demonstrates that the physics associated with cognitive science is not a matter-based as it might seem.
18. "New Materialism," *Sydney College of the Arts, The University of Sydney*, accessed August 25, 2017, http://sydney.edu.au/sca/research/new-materialism/index.shtml.
19. See Bruno Latour, *Reassembling the Social: An Introduction to Actor-Network-Theory* (Oxford: Oxford UP, 2005).

20. See, for example, Jane Bennett, *Vibrant Matter: A Political Ecology of Things* (Durham: Duke UP, 2010), 17–19, and Alfred Jarry, "Clinamen", in *Exploits and Opinions of Dr. Faustroll, Pataphysician*, trans. Simon Watson Taylor (Boston: Exact Change, 1996), 88–94.
21. As Ian Bogost characterizes it, "Object-oriented ontology ("OOO" for short) puts *things* at the center of this study. Its proponents contend that nothing has special status, but that everything exists equally—plumbers, cotton, bonobos, DVD players, and sandstone, for example. In contemporary thought, things are usually taken either as the aggregation of ever smaller bits (scientific naturalism) or as constructions of human behavior and society (social relativism). OOO steers a path between the two, drawing attention to things at all scales (from atoms to alpacas, bits to blinis), and pondering their nature and relations with one another as much with ourselves." "Object Oriented Ontology," accessed August 25, 2017, http://ooo.gatech.edu.
22. Tellingly, metaphors deployed to describe the operations of DNA in the 1980s and 1990s are primarily linguistic metaphors, although Francis Crick originally erred in denominating DNA a "code" instead of a "cipher." But Crick's mistake was the first symptom of the grand shift already taking place in the relation between human subjects and the Law. From the late 1990s to the present, digital metaphors such as "information" have taken over, showing the link between digital and machine technologies and thinking about biological processes. There is no real here or increased understanding, merely evidence of a cultural shift to the machine which merely stages out and dilutes the operation of anything like Law.

4

"A Corporal Radioscopy": Lacan, the Baroque, and the Posthuman

Allan Pero

In *Encore*, Lacan insists that "I am situated essentially on the side of the baroque."[1] What does this situation imply? Certainly, Lacan has been both praised and castigated for his "baroque" style of writing and presentation, but, if we are going to take his declaration seriously, what are the implications of this remark, beyond those of attempting to speak and write in a way that remains faithful to the subtlety and complexity of mathemic thought? In order to answer this question, I want to follow Lacan's own advice and take him *literally*. To situate himself "on the side of the baroque" implies adopting a perspective, a space, which has been vouchsafed him by its historical and cultural beneficence. This attitude should not surprise us, given how important a role the thought and culture of the baroque plays in the development of Lacan's work. Of course, the baroque is notorious for being marked by a fascination with theatricality, the bloated, even anamorphic distortion of the

A. Pero (✉)
Department of English and Writing Studies, Centre for the Study of Theory and Criticism, University of Western Ontario, London, ON, Canada
e-mail: apero@uwo.ca

S. Matviyenko and J. Roof (eds.), *Lacan and the Posthuman*, The Palgrave Lacan Series,
https://doi.org/10.1007/978-3-319-76327-9_4

unity and precision of the High Renaissance, the seventeenth-century crisis of authority, and is conjoined with a Counter-Reformation politics and religion. Although Lacan had renounced his Catholicism, it must be admitted that Catholic and Christian motifs and logics persist in his work; indeed, the Rome discourse on the symbolic, real, and imaginary, and their accompanying Borromean knot, is simultaneously Christian and baroque in its configuration. The unholy trinity of the knot, borrowed from the famous baroque family, Borromeo, was first appropriated first by them from the Christian iconography of the Middle Ages. As you will recall, the Borromean knot is not properly a knot, but a kind of chain or topological machine, in which severing one of the knots prompts the severing of all three; there is a unity inherent to the trinity such that each is dependent upon the other in order for each knot to retain its integrity.

The Borromean knot can this be understood not as the Lacanian subject, but as the spatial structure of the Lacanian subject's relation to the symptom. It is not an illustration, but a topological machine, producing the relations between and among the symbolic, the real, and the imaginary. The materiality of the structure, which would seem to situate it more properly in humanism, is, I will suggest, an index of Lacan's thought being posthuman *avant la lettre*. The Borromean knot, coupled with Lacan's other adventures in topology, works together to radically re-conceptualize *jouissance*. If Protestantism is a privileging of the subject's symbolic access to the "real," of direct communion with God without mediation, then *jouissance*, as a concept, *suffers*: The real is reduced to yet another site of knowledge, and *jouissance* becomes limited to the purely corporeal, phallic, finite experience of orgasm. But Lacan's interest in the baroque, in its re-assertion of the other *jouissance*, is governed precisely by the mysterious, enigmatic relation to the signifier, and not by the fantasy that one can generate a relation to the real in clear, plain speech or "dialogue." This relation, Lacan contends, is always structured by "passion"—that is, he is attempting to get us to re-imagine our passion for the effects the signifying function has. Why does he use this particular word—"passion"? The word has several meanings, some of which are in play here: The first is in the lay sense of strong emotion.

For example, our resistance and transference to Lacan is an index of our very emotional relationship to language. A second meaning of passion is endurance, an enduring of mysterious suffering—of the kind that Christ endured on the Cross.

But what prompts this passion, this mysterious identification? The effect or function of signification itself—the machinic signifier—what Lacan will later call the symptom or *objet petit a*—which assigns you a particular place in the Symbolic network; it insists itself upon you, re-inscribing itself in your life. Because our relationship to the letter is unconscious, the letter as symptom remains, in some sense, mysterious to us (like a foreign tongue). The signifier as machine constitutes the split between thinking and being, even as it produces the subject. Self-consciousness, or the recognition of the unconscious, necessarily means that we give up the fantasy of simply "being" as soon as we start thinking about what "being" is. Thus, Lacan emphasizes the baroque dimension of Descartes's thought by recasting the aphorism "I think, therefore I am" as "I think where I am not; therefore, I am where I do not think." Where "we are" is predicated on the radical dislocation on which this chiasmus depends. In this formulation, we are asked to think about a concept that, properly considered, is paradoxical. How can I think the deliberate asymmetry that this sentence demands if some element of it is outside the field of thinking? What Lacan's "baroquism" asks us *to feel* is that we are located always in a hidden place.[2] That is, it appears in the place where, as Lacan puts it, we "do not think to think." This passion for the effect of the signifying function is the speaking "it" to which Lacan refers (*ça parle*); in other words, it is our relation to the signifying function, to the "it" of speech that provokes our passion, because "it is not only man who speaks, but that in man and through man it speaks (*ça parle*)," and "that his nature is woven by effects in which is to be found the structure of language, of which he becomes the material, and that therefore resounds in him ... the relation of speech."[3]

You might have caught the metaphor of penetration that Lacan uses to describe how the subject is "shot through" by the signifier; here, we discover one of the ways in which the passion of suffering the signifier is narrated in his thought—as *jouissance*. In a sense, a state of ecstasy, as experience, becomes a kind of knowledge, but knowledge of a special type; that is, it cannot be reduced to "trivia." It is the knowledge caused by what the Lacan calls "*jouissance*"—which not only means "orgasm," but also refers to the "painful pleasure" or "passion" that comes with "enjoyment" or "bliss." Let us think about enjoyment or bliss. This particular kind of enjoyment always implies that there is something about the experience of blissful enjoyment that somehow defies internalization. There is always something leftover, some kind of materiality, residue or remainder which renders the experience senseless, as an indigestible "something" without meaning. This excess is usually conceived as an index of phallic *jouissance*; I want to consider more fully why it is, in a different sense, part of the province of the baroque's privileging of an experience which has nothing to do with the human body.

As we see, the question of ecstasy is a complicated one, both sexual and religious; indeed, it is a truism in baroque art that several of the sculptures and paintings by artists like Giovanni Benedetto Castiglione, or Gianlorenzo and Giovanni Bernini—for example, the famous sculpture of St. Teresa of Avila (1652), which graces the cover of the French edition of *Encore*—in their depiction of religious ecstasy, look startlingly similar to the body in a state of sexual ecstasy. But, as Lacan reminds us, they are not the same experience (I will discuss this distinction in more detail below). Rather, religious *jouissance* is more a state than a mere feeling. But what is "ecstasy"? Ecstasy is a displacement; the person is rendered "out of place." But, for Lacan, the baroque has much more to teach us about ecstasy than sloppily conflating one form of *jouissance* with another; this is one of the reasons why he is situated on the side of baroque. The baroque explores and re-constitutes the relation of jouissance to excess. In its advocacy of the *jouissance* of the other, the baroque is supposed to tarry with the borders between what can be symbolized and what is abject, excessive, or in Lacanian terms, the real. If the baroque, as Jorge-Luis Borges argues in the preface to his *Universal History of Iniquity*, is "that style that deliberately exhausts (or tries to exhaust) its own possibilities, and that borders on self-caricature,"[4] what are the possibilities its attempts to exhaust? I would contend, along with Christine Buci-Glucksmann, that Lacan's interest in the baroque stems from its playful exploration of "the *infinite materiality* of bodies and images."[5] This infinite materiality suggests that there is something alienated in/from the process of symbolization, that, the repetition and excess that characterize the baroque are themselves indices of a general aesthetic of "incompleteness" as José Antonio Maravall describes the period.[6] That is, each aesthetic object attempts to go to an extreme that, in reaching it, points paradoxically to a beyond or outside that it has failed to capture; for Lacan, that extreme is not bodily or phallic *jouissance*, but rather the Other or feminine *jouissance*. An aesthetic of incompleteness suggests that one of the baroque's governing principles cannot be self-identity; as Lacan himself says, part of his being "situated on the side of the baroque" is an effect of unconscious enunciation; he asks: Do I situate myself? Or am I situated by the Other? Of course, the answer is: "Yes, and/but no." This effect is, for him, a marker of the "subtlety of lalangue" (his word for those elements of language that resist meaning, and play with ambiguity and puns). This subtlety is the very problem that the baroque doctrines of extremism and incompleteness are attempting to capture. In other words, the subtlety of the baroque is not to be found in its theatricality or its excess, but in the puzzling mechanics and affects produced by the substance of *jouissance*. In order to think through the implications of this contention, we must turn away momentarily from the baroque and consider the topological and cybernetic dimensions of enjoyment.

The substance of *jouissance*, be it phallic or that of the Other, appears at two conjunctions of the Borromean knot: The first occurs in the topological zone between the symbolic and the real; the second occurs in the zone between the imaginary and the real. The suzerainty of phallic *jouissance* is, paradoxically, produced by its finitude. It is "detectable" as a limit to the symbolic and is thus comprehended only as the moment language and law fail or cease to function; phallic *jouissance* is thus understood as the pleasure principle. But no such understanding is possible for the other's *jouissance* because it is without symbolic content; in its joining of the real with the imaginary, the other *jouissance* can be experienced only as the limit of fantasy's collusion with impossibility. Since the imaginary can only be known retroactive to one's movement from homeomorphic to heteromorphic identification, then it lacks the requisite tools to articulate this experience as a symbolic limit, and can be recognized only as a fantasmic one. From a topological standpoint, homeomorphic identification makes it impossible to distinguish the subject from the other, or the subject from its image, because it is, properly speaking, isomorphic. The mirror stage introduces a radical cut in the set of isomorphic features which make up the imaginary identification, and thus prompts a shift in identification whereby the emerging subject can distinguish itself from the other. The latter form of enjoyment, articulated with the matheme *JA*, is not registered bodily; it is, I will suggest, registered cybernetically—or, more specifically, in a posthuman form.

If the symbolic order generates our idea of the universal, and indeed, constitutes itself as such (that is to say, that from the perspective of the symbolic, there is no universe outside that which can be symbolized), then the differentiation between subject and other which inheres in symbolic identification ultimately encounters its own scotoma or blind spot. Differentiation between the symbolic and the real ultimately plays itself out on the field of universalism, of "a symbolism which they [researchers] bring into operation inside the real, neither by virtue of projection, nor as a framework of thought, but by virtue of being an instrument of investigation."[7] Another way of putting it is to say that the symbolic, in positing itself as universal, ultimately engages with the real narcissistically. The differentiation occurs, as Lacan insists, "in the seam where the imaginary joins the real."[8] Although he cautions us to remember that the imaginary can often be confused with the real, it is only in this field that the differentiation can be experienced. If we, in our symbolic attempts to identify with *objet a*, that little bit of the real, our attempts perforce founder on our universalism, on our insistence that the symbolism *is* the universe. If fantasy, in its relation to the other, is an image set to work to try answer the question of the other's desire, then the imaginary confronts another problem; it sees the real as an object of desire, but it at the same time sees the real as it is, without fissure, unsullied by the lens of symbolism, and it is quite impossible to "give" to someone else.[9] In effect, the symbolic sees the universe reflecting back only itself, whereas the imaginary sees that the universe, and its unfissured presence, reflects only the limits of fantasy's ability to conceive, much less answer, the real of desire. Yet Lacan later avers in his seminar on the baroque, that the truth, at least in its guise as a "dit-mension" of the real, cannot be spoken of any better than in the Gospels, whose effect he compares to "back[ing] reality into fantasy."[10] In this sense, then, one can say that the appearance of truth, is cloaked, for Lacan, in fantasmic, not symbolic clothing.

The notion that truth emerges in the seam between the imaginary and the real dovetails into Lacan's repeated claim that language "isn't a code."[11] In this respect, Katherine Hayles's *How We Became Posthuman* neglects to consider the nuance of Lacan's contention, when she argues for the complexity of word processing codes over his view of language. She compares his reading of "floating signifiers" to her notion of "flickering signifiers" and imagines that, through information technologies, she has found a way out of the impasse of what she calls his "psycholinguistics." I will quote her at length:

> Lacan, operating within a view of language that was primarily print-based rather than electronically mediated, not surprisingly focused on presence and absence as the dialectic of interest. When he formulated the concept of floating signifier, he drew on Saussure's idea that signifiers are defined by networks of relational differences between themselves rather than by their relation to signifieds. He complicated this picture by maintaining that signifieds do not exist in themselves, except insofar as they are produced by signifiers. He imagined an ungraspable flow floating beneath the network of signifiers, a network that itself is constituted through continual slippages and displacements. Thus, for him, a doubly reinforced absence is at the core of signification–the absence of signifieds as things-in-themselves as well as the stable correspondences between signifiers. The catastrophe in psycholinguistic development corresponding to this absence in signification is castration, the moment when the (male) subject symbolically confronts the realization that subjectivity, like language, is founded on absence.[12]

This quotation requires some unpacking. When Hayles argues that Lacan thinks primarily of language (or signifiers) in terms of print media, she has made a foundational error. Lacan's theory of language and signification bases its materiality in its effects on the subject, but is only incidentally concerned with print-based media. Rather, Lacan follows Saussure in viewing the signifier as a primarily phonological, mental image, rather than as a visually determined symbol. Although Hayles is correct in assuming that signifiers are marked by relational difference, and that the subject's relation to the Master signifier is structured around absence, she is mistaken in her "negative" reading of this absence. That is to say, it is precisely this absence which makes several productive effects possible: the development of the subject of the unconscious; a yielding of the fantasy of being the phallus; a symbolic identification with the other; the emergence of the engine of desire; and the rich, differential relationship of signifiers to each other.

Hayles goes on to articulate her final contention with regard to the psychology of the posthuman subject's relation to information; the emphasis shifts from presence/absence (as in Lacan) to pattern/randomness—a binary that places the subject at the computer in "command" of input, once the "mutation" of randomness and contingency generates its own "catastrophe" (one she sees as analogous to castration) in the realm of pattern. Ultimately, the effect of flickering signification for the posthuman resides in its shift from the "Lacanian … generative coupling of linguistics and sexuality" to "the fascinating and troubling coupling of language and machine."[13] The problem that Hayles claims to have solved has, in fact, simply been re-iterated. In *Seminar II*, Lacan shows us that he already understood language as a technology shaping the subject's relation to the symbolic order and the real. The question, then, is not what distinguishes the human from the machine; the question, Lacan contends, is "knowing whether the human, in the sense you understand it, is as human as all that."[14] The pattern/randomness binary which Hayles seems to privilege has the milieu of the posthuman is endemic to the language machine. The signifier is not only in a differential relationship to other signifiers, it is also structured by a signifying chain that is governed both by metaphoric, random substitution (a compensatory stratagem for the repressed object of desire and castration that dissolves the Oedipus complex) and by the pattern of displacement, or desire, which keeps the signifier from maintaining a fixed connection to the ever-sliding signified.

The machinic conception of language, which Lacan explores in his lectures on cybernetics in the 1950s, revolves upon the problem of the real in language's invocation of law; as he is at pains to point out, prior to the exact sciences, "man" did not "pretend to lay down the law, he pretended to be indispensable to the permanence of the law. An important definition, for in truth it entirely safeguards the rigour of the existence of the real."[15] The rise of the exact sciences is thus not a threat to the real, but to our relation to the real. The patterns which inhere in ritual—invoking the Gods of the sun or rain, for example—were meant to keep the real in its place, and protect humanity from the vicissitudes of randomness. In effect, the exact sciences are in danger of reducing the real to the kind of symbolic universalism that I discussed earlier. But the history of this advance, which would of course include cybernetics, is much longer than posthuman critics like Hayles would have us believe. The posthuman, at least in its relation to pattern and randomness, is ultimately an invention of the baroque. Lacan reminds us that the development of probability calculus by Pascal, whose treatise on the arithmetic triangle (a figure Lacan calls "the first machine") appears first in 1654, is a means of organizing contingency—a symbolic assault on our traditional conception of the real—through a series of probable patterns of presence and absence.[16] Rather than narrate the history of psycholinguistics as a shift from absence/presence to pattern/randomness, as Hayles does, Lacan sees the cybernetic logic of presence/absence as working *in tandem* with that of pattern/randomness to generate a discourse in which the real is an inert wall, passively awaiting the installation of doorways through symbolization.

I use the image of door and wall advisedly; Lacan himself uses it to explain the limit placed on the door in its incursion on the wall of the real. He asks us to "imagine a great wall which would go around exactly round the earth" and poses the question "if you knock a door through it, which is the inside, which is the outside?".[17] The door opens and closes upon a circuit, which he conceives of as a wall, on what does it open? The real or the imaginary? We cannot say, Lacan avers. If we construct a machine that opens and closes doors as handily as it might a circuit, have we then produced a machine that thinks? Lacan's answer is that we have not, any more than the person who opens or closes the door unthinkingly.[18] There is a mechanical dimension to human action, to human desire, just as there is to the feedback in a circuit.

Yet another reason why we cannot determine which side of the door is the inside or the outside is that the wall or circuit is itself analogous to the effect of the Moebius strip; the binary of presence/absence that Hayles rightly sees in Lacan's theory of language disregards that the topology of Moebius strip is deployed by Lacan for the purpose of troubling the binary's aridity. What he attempts to show us is that, like the strip, notions like inside/outside, or presence/absence, are not discrete terms, but are in fact continuous with each other. The sides are continuous, but there is only one surface. It thus becomes impossible to determine the place at which one has crossed from the inside to the outside, or from the outside to the inside, once the circuit (or door, if you prefer) is closed. The Moebius effect also applies to the binary human/machine; Lacan's materialist conception of consciousness (which is linked to his theory of language and the unconscious) disrupts the dualism which inheres in Hayles' posthuman model, which persists in anthropomorphizing the machine, even as it exalts the category of cyborg. As Louis Armand writes in his article "Symptom in the Machine: Lacan, Joyce, Sollers,"

> The machine does not take the place of consciousness (or of conscience), nor does it absolve the subject of the burden of the "Real." Rather it situates the subject as a figure of recursion (a mechanism, in fact, of recursive substitution), simultaneously inscribed within and exceeding the symbolic organisation of the machine.[19]

Armand proceeds to explain the implications by demonstrating that the ego is perforce a function of the imaginary and thus has no symbolic dimension, except insofar as the symbolic "assigns to itself an implicitly mechanical placement in the circuit of symbolic exchange." If we follow the logic set out here to its conclusion, we are confronted with two notions: (1) that the symbolic order is itself a machine; and (2) that when the ego is divested of its mechanical, symbolic dimension (that is, the illusion of unity), we are better able to see the ways in which the symbolic machine "returns" desire to the subject. The symbolic, as machine, produces the desiring subject, and the subject is "therefore its source of nourishment."[20]

But what role, then, does unconscious fantasy play in this circuit? As I have already suggested, Lacan sees it "as an image set to work in a signifying structure."[21] Fantasy functions as a kind of valve, regulating the subject's unconscious relation to desire such that aphanisis, the vanishing of desire, does not cause the neurotic subject to disappear completely when demand is satisfied by the signifier—the object of desire. In conceptual terms, aphanisis is linked to the binary presence/absence, insofar as the satisfaction of demand, which prompts the vanishing of desire, points negatively back to the emptiness of the subject without desire. That is to say, when the subject's desire is, however momentarily, satisfied or represented the object or signifier, the gap between the emptiness of the subject and the richness of its symbolic narrative is made plain. Even as the neurotic subject longs for the fading of desire, fantasy does not ultimately permit utter fading to occur; "it is the support of desire; it is not the object that is the support of desire."[22] We might call this the "alienation effect" of desire; as Lacan tells us, "There is no subject, without, somewhere, *aphanisis* of the subject, and it is in this alienation, in this fundamental division, that the dialectic of the subject is established."[23] The anxiety of alienation paradoxically sustains the subject in the moment of fading. The subject fades, but does not disappear. The reason is that the unconscious, being structured in an extimate relation between subject and other, is "the discourse of the circuit in which [the subject] is integrated."[24] The subject is thus not so much marked by personal traits, as it is by its position in the circuit.

André Nusselder reads Lacan's excursion into cybernetic thought as crucial to understanding fantasy. For Nusselder, the subject is not merely an effect within in the symbolic network; it is *virtual*. Fantasy is a regulating force, "framing our perception" of our position in the symbolic circuit, mediating between the ways in which the subject engages with signifiers, and the ways in which the symbolic order reads or responds to the subject's use of fantasy and language. Our identity is to be distinguished from our place in the symbolic circuit. If identification with the symbolic is the means by which we are inscribed in the circuit, then our continued fantasmic identification with images is that which constitutes our "personalities."[25] Most importantly, "fantasy is hence at the interface of the real and the virtual."[26]

If we take up Nusselder's useful and provocative insight, what are its implications to Lacan's own identification with the baroque? First, we must consider how Lacan thinks about the baroque, both as a period, and as a narrative. He suggests that the "baroque is, at the outset, the 'storyette' [*historiole* in the French text] or little tale of Christ."[27] What is crucial to recall, Lacan goes on to argue, is that this baroque tale "is presented not as an enterprise of saving men, but as that of saving God."[28] If Christ's sacrifice is narrated by the Gospels as the salvation of humanity, then what is the baroque's function—since Lacan understandably distinguishes it from scripture? Saving God, it seems. But from what? The answer would appear to be God's incorporation—as the real—into the symbolic. The baroque, then, has a fantasmic function as well as a symbolic one. If, at the level of the symbolic, the baroque is working to produce a Moebius-like torsion of the story of Christianity in response to the rise of Protestantism, it is also contending with fantasmically protecting the Other's *jouissance*. Elsewhere in *Encore*, Lacan works to clarify what he means by God; although he has been understood (precipitately) as having helped to eliminate God from the discussion of existence (that is, if God is dead, then everything is prohibited), he is at pains to demonstrate "in what sense the good old God exists."[29] God exists, for Lacan, not for theological reasons, but because he needs the concept of God in order to save the Other—for the subject.

This is part and parcel of the paradox of *jouissance*: it is the surplus of enjoyment that is derived from the symptom, even as one suffers as a result of it. It is, in other words, the satisfaction that comes from suffering for your desire. It would seem then, that the removal of the obstacle to enjoyment—the law of the father—would make more enjoyment (as pleasure possible). But as Lacan insists, following Freud's *Totem and Taboo*, the murder of the father does not "open the path to jouissance," but instead "strengthens the prohibition."[30] The act of murdering the father becomes not a means of ensuring more satisfaction, but accruing a debt that is inscribed in the symbolic order. As a result, the superego becomes all the more relentless the more one tries to submit to its edicts. Prohibition, then, becomes crucial to enjoyment in that enjoyment itself fashions a trench or track for the drive; prohibition is thus the condition of enjoyment. This is the link between desire and the law; desire is produced by the prohibitions set up by the law—the name of a father who exists, but is always already dead. This is why the symbolic father—the father of the law—is always a dead father. Think of Nietzsche—"God is dead insofar as we have murdered him." It's not a "turning away from God" of which Nietzsche has been so often accused. The graffito "God is Dead—Nietzsche; Nietzsche is Dead—God" utterly misses the point, and is, by its own lights, blasphemous, because it appropriates the voice of God in order to give us the rather banal news that Nietzsche is dead. Why is the father dead? God dies in order to start the chain of signification going. To unite the desire and the law through love; love of the dead father is the symbolic debt owed to him. Christ's passion is the opportunity afforded Catholics to love God, to make his resurrection possible through the act of love—this is one effect of baroque cybernetics.

The existence of the Other—God—is thus essential both to thinking and being. If there is a fundamental split between thinking and being, which, following his chiasmic twisting of Cartesianism, is necessary to the development of the cogito as subject of the unconscious, then there must be an Other for the unconscious to exist as an ambivalent gift of love, and *jouissance* to think itself into being.[31] If thinking, for Lacan, is the "jouissance of being," then he must hold on to the concept of the Other *jouissance* in order to situate the role of Woman's *jouissance* as distinct from that of the phallus. The coincidence of Woman as symptom (*objet a*) and the Other *jouissance* occurs at the pleasure principle.[32] In effect, this coincidence is the means by which *jouissance* can be "known" in the symbolic, in the same way the Woman as symptom of man negatively defines what constitutes man as subject. Here, we find the conjunction of three elements of machinic desire: fantasy, *objet a*, and phallic *jouissance*; the conjunction of Woman as symptom and the Other *jouissance* is necessary for the two forms of *jouissance*—phallic and that of the Other—to exist. The former is experienced as an identification with the "masculine" side of the formulas of sexuation, while the latter is, as you will recall, the result of its position in the Borromean chain, the site of "pure" *jouissance*. The differential, but mechanically connected, types of *jouissance* also preclude the possibility of a sexual relationship. Phallic *jouissance* is not the pledge and seal of a sexual relationship because it is connected only tangentially, but not directly, to the Other *jouissance*; with respect to the Borromean chain, they are dependent upon each other. In other words, there can only be a sexual relationship if *jouissance* no longer exists.[33] Moreover, the only way for a sexual relationship to exist is in sadistic or perverse fantasy; in it, the identification with phallic *jouissance* is radically confused with *objet a*; the subject assumes the position of *objet a*, foreclosing the *nom-du-père*, and stages the real of castration by making the other literally experience the split in her subjectivity.

With these elements in place, we can now turn back to the baroque, and its role in the development of the posthuman. The effect of the baroque mechanism—in aesthetic terms—is to stage "the body evoking jouissance … but without copulation."[34] Uncannily, this very impulse drives much scientific thought in the baroque era. For example, the mathematical and technological advances in optics and lenses in the seventeenth century, as Ofer Gal and Raz Chen-Morris describe it, "mediat[e] between fantasy and truth" in a manner that, rather than produce a direct "perception of nature, … suggested a path to a direct encounter with the divine. Fascination with mirrors and lenses reversed the Pauline dictum: Only the enigmatic mirror, by submerging oneself in its reflections and apparitions, could one find oneself "face-to-face" with one's creator."[35] The baroque is thus not simply an "over-the-top" theatricalization of religious ecstasy; rather, it is a way of bringing the real into the scopic "dit-mension" of fantasy for the purpose of generating truth. In terms of Lacan's defense of baroque science, fantasy must be brought into the discussion of the real in order to undercut modern science's fatal conflation of its discourse with truth. The way it is managed is through obscenity.[36] But what does Lacan mean here? He is not simply suggesting that the baroque is perforce "obscene" by refusing to acknowledge its conflation of phallic and Other *jouissance*—this would be to miss the point entirely. For the baroque, the body-as-machine has an allegorical function; it makes visible the obscenity of phallic *jouissance* by focusing negatively on a different kind copulation—a non-sexual copulation—of the real and the virtual to produce the "entelechy of this body": What Lacan, in his baroque re-reading of Aristotle, calls the soul.[37] The product of this kind of copulation, unlike the senselessness or blasphemies of phallic *jouissance*, is *thinking*. The soul, in this formulation, is the thing that thinks; the baroque thus stages a thinking of the *jouissance* of being. It is for this reason that when Lacan exhorts us to look at Bernini's Saint Teresa, he can make a delicate, important distinction in the midst of vulgarity; the viewer "immediately understands that she's coming. What is she getting off on? It is clear that the essential testimony of the mystics consists in saying that they experience it, but know nothing about it."[38] Thinking is, in the Other's *jouissance*, detached from the phallic desire to know, and is further, in its collusion of the virtual and the real, the engine of entelechy's circuit. The "experience" to which he refers is the baroque's anticipation of posthumanism's critique of the human as the apex of existence, and of the fundamental unity of the human individual; the baroque conception of mystical experience is the incorporation of the real through the soul, or through the circuit produced by what Nussfelder calls the "interface" of the virtual and the real. As the golden metal beams surrounding, indeed, radiating into Bernini's Saint Teresa suggest, the result is a cybernetics of *jouissance*, staging the monstrance of Saint Teresa's accidental body, and the presence of the real in the soul's entelechic ecstasy. It is an experience of separation, of the ecstatic, rather than the experience of alienation produced by the subject's relation to the signifier.[39]

However, the problem Lacan encounters in conceiving the Other *jouissance* as JA (which he does not resolve in *Encore*) is that it runs counter to his repeated insistence elsewhere of the lack in the Other—that there is no Other to the Other. If S(A) obtains, then Lacan must acknowledge that the matheme JA cannot. There are several reasons for a necessary shift in Lacan's late thought; in the final incarnation of the Borromean chain, which makes its appearance in *Seminar XIII*, the interface of the imaginary (the virtual) and the real is written with the matheme J~~A~~. This shift does not mean that the Other's *jouissance* can now be conflated with that of the phallus; there must remain a pure site for the real to be encountered as One, the same impulse towards the One which, as Lacan wonders in *Encore*, is part of the tension which shapes Eros—the desire to be one with an other.[40] As Lorenzo Chiesa deftly explains in *Subjectivity and Otherness*, the distinction between JA and J~~A~~ is the following: While the former can be understood as pure *jouissance*, the latter "is therefore (a form) of *jouissance* of the impossibility of JA."[41] The purity of the real as an experience, like the symbolic's later recognition of the imaginary, "must," he suggests, "be presupposed retroactively, but it cannot be counted as (a self-enjoying divine) One, not even as the supposedly 'weaker' One of pure difference."[42]

What then, does this shift mean in terms of Lacan's fidelity to the baroque? It prompts a re-imagining of the Borromean chain through the emergence of the sinthome, even as it sustains the logic of incompleteness which characterizes the baroque. It is how barred *jouissance* (J~~A~~) makes itself experientially possible as a fourth link in the Borromean RSI machine. The sinthome is a conceptual leap from the symptom insofar as it utterly independent of the symbolic and is thus unanalyzable. It sustains the Borromean chain (which, as Lacan points out, is no longer sustained by itself, but by the sinthome, but is in no way dependent upon it).[43] In this respect, the fourth link enjoys a status that the other three do and cannot. The sinthome prevents the collapse of the subject into psychosis by producing a kind of radically Other *jouissance* which stands outside, forbidding imaginary capture, but not precluding an encounter with the imaginary; in this sense, the mystical experience which he privileges in *Encore* returns, but, as topology, functions not as "an occultation of the symbolic,"[44] but of the imaginary—more specifically, in its fantasmic encounter with the real outside the symbolic. The sinthome can be understood not simply as yet another "medium" for the prosthetic God, but as Lacan's confrontation with the limits of the symbolic to contend with the imaginary and the real. If the symptom was developed as a means of compensating for, and identifying with, a lack in the subject, then the sinthome emerges from an identification with the lack in the Other. Just as the imaginary link in the Borromean chain is the mediator between the symbolic and real, so, in its consistency, the sinthome is importantly, an *invention*—it mediates within the imaginary between and among the three registers, even as it makes it possible to distinguish them. Crucially, the sinthome has another function; just as the "baroque is the regulating of the soul by corporal radioscopy,"[45] the sinthome regulates the subject in a way which makes an identification with the J$\not{A}$ possible. Unlike the symptom, which the subject believes is an enigma with a solution, a master-signifier that will free the subject from his neurotic desire, the sinthome regulates the soul's entelechy by bringing to light—as it does with the radiant beams penetrating the cyber-soul of Bernini's Saint Teresa—a way of being that is not governed by the unconscious or its phallic vicissitudes, but without the foreclosures of the *Nom-du-père* which mark psychosis. The sinthome's particular corporal radioscopy reveals a substance lurking in the suture separating the imaginary and the real; a baroque (and yes, posthuman) machine of individuation, naming and occupying the gap in the Other *jouissance*.

Notes

1. Jacques Lacan, *The Seminar of Jacques Lacan: Book XX Encore, 1972–1973*, trans. Bruce Fink (New York: Norton, 1999), 106.
2. Lacan, *Encore*, 113.
3. Ibid., 284.
4. Jorge Luis Borges, *A Universal History of Iniquity*, trans. Andrew Hurley (Toronto: Penguin, 2001), 4.
5. Christine Buci-Glucksmann, *Baroque Reason: The Aesthetics of Modernity*, trans. Patrick Camiller (London: Sage Publications, 1994), 139.
6. José Antonio Maravall, *Culture of the Barogue: Analysis of a Historical Structure*, trans. Terry Cochran (Manchester: Manchester University Press, 1986), 209–211.
7. Jacques Lacan, *The Seminar of Jacques Lacan: Book II, The Ego in Freud's Theory and in the Technique of Psychoanalysis, 1954–1955*, trans. Sylvana Tomaselli (New York: Norton, 1988), 98.
8. Lacan, *The Ego in Freud's Theory and in the Technique of Psychoanalysis*, 98.
9. Ibid., 253–254.
10. Lacan, *Encore*, 107.

11. Lacan, *The Ego in Freud's Theory and in the Technique of Psychoanalysis*, 279.
12. N. Katherine Hayles, *How We Became Posthuman: Virtual Bodies in Cybernetics, Literature, and Informatics* (Chicago: Chicago University Press, 1999), 30–31.
13. Hayles, 35.
14. Jacques Lacan, *The Ego in Freud's Theory and in the Technique of Psychoanalysis*, 319.
15. Ibid., 297.
16. Ibid., 299.
17. Ibid., 300.
18. Ibid., 302–304.
19. Louis Armand, "Symptom in the Machine: Lacan, Joyce, Sollers," *The Symptom 3*, (2002), http://www.lacan.com/sympmach.htm.
20. Lacan, *The Ego in Freud's Theory and in the Technique of Psychoanalysis*, 54.
21. Lacan, *Écrits*, 532—Need full citation for *Ecrits* here.
22. Jacques Lacan, *The Seminar of Jacques Lacan: Book XI The Four Fundamental Concepts of Psycho-Analysis*, trans. Alan Sheridan (New York: Norton, 1981), 185.
23. Lacan, *The Four Fundamental Concepts of Psycho-Analysis*, 221.
24. Lacan, *The Ego in Freud's Theory and in the Technique of Psychoanalysis*, 89.
25. André Nusselder, *Interface Fantasy: A Lacanian Cyborg Ontology* (Cambridge: MIT Press, 2009) 76.
26. Nusselder, 75.
27. Lacan, *Encore*, 107.
28. Ibid., 108.
29. Ibid., 68.
30. Jacques Lacan, *The Seminar of Jacques Lacan: Book VII The Ethics of Psychoanalysis, 1959–1960*, trans. Dennis Porter (New York: Norton, 1992), 176.
31. Lacan, *Encore*, 70.
32. Ibid., 84.
33. Ibid., 112.
34. Ibid., 113.
35. Ofer Gal and Raz Chen-Morris, *Baroque Science* (Chicago: University of Chicago Press, 2013), 77–78.
36. Gal and Chen-Morris, 113.
37. Ibid., 88.

38. Lacan, *Encore*, 76.
39. Jacques-Alain Miller, "The Sinthome: A Mixture of Symptom and Fantasy," in *The Later Lacan: An Introduction*, eds. Veronique Voruz and Bogdan Wolf (Albany: SUNY Press, 2007), 61.
40. Lacan, *Écrits*, 5.
41. Lorenzo Chiesa, *Subjectivity and Otherness: A Philosophical Reading of Lacan* (Cambridge: MIT Press, 2007), 187.
42. Chiesa, 187.
43. Jacques Lacan, *The Seminar of Jacques Lacan XXIII: Le Sinthome, 1975–1975*, ed. Jacques-Alain Miller (Onicar?, 1976–1977), 13.
44. Lacan, *Écrits*, 368.
45. Lacan, *Encore*, 116.

5

Lacan's Cybernetic Theory of Causality: Repetition and the Unconscious in Duncan Jones' *Source Code*

Colin Wright

Posthumanism: An Inside-Out Glove

Bringing the proper name 'Lacan' and the academic field of interdisciplinary research known as 'posthumanism' together cannot be a matter of a hand fitting into a pre-tailored glove. One could *force* a fit, by enlisting a very academic Lacanian theory abstracted from its proper roots in clinical practice, into the cause of an equally academic 'posthumanism' distanced from the lived materialities of contemporary technoculture. But what would be gained thereby? On several occasions, Lacan uses the metaphor of the *inside-out* glove[1] to figure the Moebius-like topology of the relation between the symbolic and the imaginary. The point of this figure is to show how that relation is both continuous and incommensurable: thus, it is possible to wear an inside-out glove, but only on condition that it is on the *wrong* hand, and there is always a little fold

C. Wright (✉)
Department of Culture, Film & Media, The University of Nottingham, Nottingham, UK
e-mail: colin.wright@nottingham.ac.uk

S. Matviyenko and J. Roof (eds.), *Lacan and the Posthuman*, The Palgrave Lacan Series,
https://doi.org/10.1007/978-3-319-76327-9_5

or seam that ensures it never quite fits. It is with a certain lack of fit that clinical rather than academic psychoanalysis works. By definition, each analysis is an exercise in exceeding pre-existing hermeneutic circles, for both analysand and analyst. Similarly, the much-discussed 'post' of post-humanism indexes the lack of fit between contemporary experience and old parameters of the human.

In what follows, I want to argue that 'posthumanism' is not a coherent theoretical framework with a delimited object of study at all: no hand, not even a robotic one, fully fits its glove. Instead, I will define 'posthumanism' as the name of a widespread cultural *symptom* that insists precisely because there is no clear place for it in the symbolic order. As such, it allows us to read the structure of some of the fantasies dominating the social link of hyper-technologically mediated, neoliberal consumer societies. Naming the symptom is a vital part of analysis. It gives the structure, rather than the content, of the fantasy a legibility *qua* unconscious message addressed to the Other. Only in this way can we become interested in the truth of our own desire *without falling prey to its imaginary lures.* At the level of the imaginary, there is *joui-sens* or enjoy-meant because the glove invariably fits. In analysis, however, meaning must get out of hand.

I propose here to interrogate, from a clinically informed Lacanian perspective,[2] a film with ostensibly posthuman themes: Duncan Jones' 2011 box-office success, *Source Code.*[3] The film's main protagonist, Colter Stevens (played by Jake Gyllenhaal), thinks of himself as a military helicopter pilot deployed to Afghanistan, only to discover that, following his 'death' in battle, he has become a hybrid composition of corpse and a computer programme called Source Code. Its purpose is to gather data from the brains of the recently dead by exploring the last eight minutes of memory 'cached' in their neural pathways. On this, his first mission, Stevens is drafted into the War on Terror: a bomb has exploded on a train, killing everyone aboard. As a string of binary code, he is 'jacked in' to one of the deceased passengers to determine who planted the bomb, and where and when the next attack will be. In *Groundhog-Day*-meets-*Mission-Impossible* fashion, the film takes us through Stevens' repeated attempts to identify the bomber by exploring the same eight-minute sequence from every conceivable angle. At the level of plot then, *Source Code* is clearly the proverbial race against time: its tagline is 'Change the Past. Save the Future'. However, I will argue that it reveals a deeper and more contradictory relationship to time and causality peculiar to the posthuman era. My main argument will be that Lacan's theory of psychoanalytic causality, developed through a critical engagement with cybernetics and game theory, can shed much light on this ambivalent posthuman temporality, as well as *Source Code*'s ultimate failure to confront it.

By approaching this film as a posthuman symptom rather than as a fantasy, I aim to contribute to *critical* posthumanism as distinct from the celebratory version associated with Hans Moravec,[4] or its scare-mongering counterpart in the paranoid posthumanism of someone like Francis Fukuyama.[5] This attempt to perform a 'symptomatic reading' of *Source Code* should be seen as part of the interpretation of the present in which both critical posthumanism and psychoanalysis are engaged, though without necessarily fitting seamlessly together, hand in glove.

Lacan's Cybernetic Theory of Causality

It is widely recognised that the algorithmic calculation of probabilities was a seminal issue for cybernetics, which accounts both for its origins in the military brinkmanship of the Cold War and its early uptake in areas such as economics.[6] Yet it is less frequently acknowledged that Jacques Lacan's work has a parallel concern with developing *a psychoanalytic theory of causality*, one sufficiently distinct from older philosophical frameworks to formalise the speaking subject's determination by the logic of the signifier—and moreover, that he was influenced in this endeavour by cybernetics. As Lydia Liu's work has shown, however,[7] there are numerous writings by Lacan that demonstrate this interest in cybernetics and causality. Here, I will limit myself to three: *Seminars II*[8] and *XI*,[9] and the *écrit* entitled 'Logical Time and the Assertion of Anticipated Certainty'.[10]

Lacan's most explicit engagement with cybernetics and game theory comes in *Seminar II* (1954–1955). It should therefore be seen within his broader attempt there to re-situate the importance of Freud's

Entwurf (originally written in correspondence with Wilhelm Fliess in 1895), which already utilised a pseudo-cybernetic approach to the neuronal system of the 'psychic apparatus' but is generally dismissed within orthodox Freudian circles as a pre-psychoanalytic failure. In *Seminar II*, Lacan is keen to show the continuities between the *Entwurf* and Freud's much later insistence on the connections between the repetition compulsion and the death-drive in *Beyond the Pleasure Principle* of 1920. This text was scandalous even to Freud's followers because it implied a form of satisfaction distinct from the facilitation/frustration of biological instincts. Lacan therefore sees in the *Entwurf* a prefigurative intuition on Freud's part of the fundamentally *machinic and therefore unnatural* status of the symbolic order that he, Lacan, would bring out more clearly with the help of Saussurean linguistics, and the notion he extrapolates from it of the unconscious as the discourse of the Other.

Beyond identifying the influence of nineteenth-century mechanistic thinking, particularly thermodynamics, on Freud's early homeostatic account of the psychic system, Lacan astutely recognises the link between his own structuralist re-reading of the Freudian unconscious, and what were then cutting-edge developments in the field of cybernetics and information theory. As Katherine N. Hayles has argued,[11] it was in this field that the concept of 'entropy' began to be reconceived in ways that left classical thermodynamics behind, suggesting to Lacan a means by which to re-conceptualise repetition in fidelity to the late Freud. Though as Liu points out,[12] he keeps his cybernetic cards close to his chest, it seems probable that Lacan was aware of the issues raised during the Macy Conferences held between 1946 and 1953, and perhaps even acquainted with the more anthropologically, philosophically and psychologically oriented work of Macy participants such as Norbert Wiener,[13] Warren McCulloch, Walter Pitts and Gregory Bateson.

This concern with causality crystallises ten years later, in the key eleventh seminar of Lacan's teaching, entitled *The Four Fundamental Concepts of Psychoanalysis* (1964–1965). In a way that has been explored in some detail by Danielle Eleb,[14] Lacan undertakes in *Seminar XI* a revision of the Aristotelian theory of causes outlined in the latter's *Metaphysics*, focussing particularly on the difference between chance and fate, with reference to Kierkegaard on repetition. As always, his concerns are clinical: How does the analysand misrecognise castration as an accident that has befallen them? And how can speech be utilised to show the analysand's complicity in the symptomatic repetition of this accident's assumed consequences? Session 5 of *Seminar XI* parses out a crucial distinction between two kinds of causality which relate to two understandings of repetition: 'automaton' refers to a kind of machinic unfolding familiar today in the idea of a programme that runs and produces predictable results, whereas 'tuché' corresponds to an unpredictable encounter with the real. Though a necessarily *missed* encounter, tuché would be an experience (such as the spontaneous production of a revealing equivocation within the analytic setting) with transformative implications that the subject can assume as their own. Something of this collision between chance and necessity in the locus of speech was at stake from the very beginnings of psychoanalysis. Freud's 'fundamental rule' of free and uninhibited verbal association was always conceived as a mechanism for producing chance at the level of the signifier. By means of this mechanism, the neurotic subject can be led to recognise, in the surprising products of their own speech, the contingency covertly at work in their experience of an oppressive necessity. To isolate, in one's supposed misfortune, the element of unconscious subjective choice, is to shift the automatic repetition of fate, traditionally meted out at the whim of the gods, towards the tychic repetition of a *destiny* that can be assumed by man in defiance of the gods. This has everything to do with the drives and arguably Lacan's most original contribution to psychoanalytic theory, the famous *objet petit a*. When he refers to this as the 'object cause of desire',[15] he has already re-formulated the Aristotelian understanding of cause into a recursive or retrospective causality grounded in a constitutive lack.

However, to fully comprehend this, I would argue that one has to read *Seminar XI* through the lens of *Seminar II*. His turn there to cybernetics is prompted by the game of 'odds and evens' that appears in Edgar Allen Poe's short story, 'The Purloined Letter'. His reading of this has been the focus of much attention in deconstructive and literary critical circles,[16] as well as, I would argue, of misunderstandings—even on the part of so sensitive a reader as Jacques Derrida.[17] Yet situating it in the context of the *cybernetic* conception of language with which the wider seminar is engaged goes a long way towards clearing up the confusion surrounding the claim to which Derrida takes exception: 'the letter always arrives at its destination'.[18] Just as information is distinct from message in cybernetics, so 'letter' in this formulation is very strongly distinguished from meaning: no-one in Poe's story reads the contents of the titular epistle, which is why Lacan uses it as an extended analogy of our determination by the opacity of the signifier. This is the case whether we understand meaning as the transmission of an intention between addressor and addressee, or, at a meta-level, as the psychoanalytic 'truth' of the link between femininity and castration that Derrida accuses Lacan of seeing everywhere. The claim that 'the letter always arrives at its destination' has to do not with an Oedipal guarantee of psychoanalytic hermeneutics, but with repetition in the unconscious conceived in cybernetic terms, as a self-regulating logical system that *has nothing directly to do with meaning though it provides its indispensable support.* If the central cybernetic insight into information is that 'meaningful' patterns are progressively constructed, via feedback loops, *out* of randomness rather than against it, it is little wonder that Lacan appealed to it, given the related primacy he grants to the 'materiality of the signifier'.[19]

Thus, by formalising Poe's game of odds and evens using pluses and minuses organised into logic tables, Lacan demonstrates that the symbolic order governs the combinatory system from which 'chance' can emerge. Chance therefore ceases to be radically aleatory, becoming instead an effect of structure. As he puts it, 'You can play heads or tails by yourself. But from the point of view of speech, you aren't playing by yourself—there is already the articulation of three signs [… which] prefigures the very meaning of the result'.[20] Because it *signifies* then, chance is that dimension of the contingent that can be appropriated by a symbolisation. When we describe something as a 'coincidence', a fragment of the real has coincided with the imaginary sufficiently to endow reality with a heightened meaning. Moreover, this symbolisation of coincidence can paradoxically be strongly *determining* for the speaking subject. It can become the unconscious cause of their desire, as when an innocuous comment in the distant past, perhaps made by a parent, takes on the properties of a master signifier ordering this subject's entire libidinal life, even leading to symptomatic effects.

Lacan's cybernetic point, therefore, is this: the possibility of 'chance' is written into the symbolic order as a mode of probability from the outset, no matter how intuitively remote this probability seems to be. To illustrate this, we could think of a lottery based on seven numbers drawn from a total of fifty. Mathematically, the probability of the sequence one to seven being picked is just as likely as a more 'random' sequence, since each time a number is selected, it has a one in fifty chance (in fact, the odds reduce as more numbers are drawn). Yet before even acquiring our ABCs, we learn to ascribe a significance to what are revealingly called *natural* numbers, and this leads us to endow this particular sequence with a disproportionate weight of meaning. Its emergence in the midst of a game of randomness is therefore experienced as a 'coincidence' of such implausibility that we are tempted to reach for the divine in order to explain it.[21]

The posthuman perspective alerts us to a danger here. Does this analogy between the symbolic unconscious and a cybernetic combinatory not threaten to support the image of the human as a mechanical input/output device, an image that fits, hand in glove, with biopolitical forms of governance and social control? Does Lacan's appeal to cybernetics paradoxically align him with the 'suturing of the subject' he identifies in modern science,[22] which is to say, with the displacement of the subject of desire by a subject-as-object of both knowledge and control? In fact, in two texts I would now like to look at in more detail, Lacan comes to our aid in fending off these very dangers.

Science's Rendez-Vous with Nature and Logical Time

In a lecture entitled 'Psychoanalysis and cybernetics, or on the nature of language',[23] Lacan acknowledges the emergence of cybernetics from the engineering problems to which Shannon and Weaver were responding with their Transmission Model of communication, when they clarified the role of signals in the encryption/decryption of a message. However, he is also clear that not only is 'message' distinct from 'code' in cybernetic informatics, but that neither of these terms is identical with *speech*.

As a faithful albeit radical Saussurean, he sees speech as irrevocably social, inter-subjective and shot through with the *imaginary*: ultimately, he insists on the impossibility of separating the 'machine' of the symbolic from its imaginary support. Indeed, he enables us to see the disembodied machine of pure information, and thus the extropian dream of the 'informational subject', as itself a compensatory fantasy. He does so in this lecture by isolating what is specific about modern science, and particularly the novel understanding of time that it develops.

The real, claims Lacan, has always been defined as 'something one always finds in the same place, whether or not one has been there'.[24] Yet he distinguishes between a pre-modern period, during which man felt that this real was fixed more securely to its place if his rites and rituals honoured it (sacrifices to sun gods, harvest festivals, fertility dances etc.), and the modern period, in which the emergence of science induces an important shift: 'from the moment man thinks that the great clock of nature turns all by itself, and continues to mark the hour even when he isn't there, the order of science is born'.[25] A completely new notion of exactitude then arises. However, it does not simply measure more precisely Newton's great 'clockwork universe', with its timeless universal laws. On the contrary, Lacan defines this new scientific exactitude as 'the encounter of two times in nature' and locates 'the source of being on time' (which is to say, science's uncanny accuracy) in 'the synchronisation of watches'.[26] Science, then, is tied 'to man's waiting'.[27] This means that at the very moment that abstract time posited itself as the pure, mechanistic measure of a law-bound universe that could be rationally mastered—from the orbit of the planets down to the 'clocking on' and 'clocking off' of industrial workers—it was taking its cue from a different time of human interests, rivalries and desires. Just as chance only signifies thanks to a preceding symbolic order, so scientific exactitude only 'counts' because *what* it counts is measured in relation to its coincidence with man's desire.

If cybernetics is valuable to Lacan, then, it is certainly *not* because it reduces the human psychic apparatus to a logical machine. On the contrary, because it avoids humanist-based technophobia, on the one hand, and extropian technophilia, on the other, the ethical stance on cybernetics outlined here is of inestimable value to *critical* posthumanism. The relevance of cybernetics lies in exactly the opposite direction to a generalisation of the human–machine analogy: 'The one thing which cybernetics clearly highlights is the radical difference between the symbolic and the imaginary'.[28] This is a statement that in no way endorses the eradication of the imaginary *by* the symbolic.[29] Indeed, it suggests the capacity of cybernetics to delineate the difference between the two in a world that increasingly produces both surplus value and surplus *jouissance* from their blurring:

> With a machine, whatever doesn't come on time simply falls by the wayside and makes no claim on anything. This is not true for man, the scansion is alive, and whatever doesn't come on time remains in suspense. This is what is involved in repression.[30]

This addition to symbolic, machinic time of a living 'scansion' is even clearer in 'Logical Time and the Assertion of Anticipated Certainty', written fully ten years before *Seminar II* in 1945. The early influence of game theory is very apparent here in the hypothetical scenario that structures Lacan's discussion. He refers to this scenario as a 'sophism' for, I would suggest, two reasons: firstly, to oppose the hyper-rationality of purely formal philosophical explanations, and secondly, to emphasise, like the sophists of old, the dimension of time known as *kairos* as opposed to *chronos*—the opportune moment as against metronomic regularity.

Although no doubt familiar to readers of Lacan, this sophistic scenario is complex enough to warrant another description here. Its key elements are as follows: three prisoners have white or black discs strapped to their backs so that they cannot see them, though they can see each other's; they are told that there are five discs in total, three white and two black; they are further told that if they can *logically* deduce what colour disc they have on their own back, they can exit through the door; and if they can then provide an account of the watertight reasoning that allowed them to be certain, they will have earned their permanent freedom.

Lacan avows this scenario's 'sinister images',[31] but they are relevant in adding to the notion of a pre-existing symbolic system a rivalrous, imaginary dimension that dramatises the extent to which subjects caught up in this system take their cue from others, who are nonetheless determined by the Other (just as the characters in Poe's story are structured in their inter-relations by the unopened letter). Nonetheless, behind the 'sinister' image of the prisoners competing with one another for freedom lies the basic logical puzzle of the a priori combinatory that consists of three white and two black discs, and their possible permutations in groupings of three. These can easily be represented graphically as (1) ●●○; (2) ●○○; or (3) ○○○.

This miniature cybernetic machine in which the inmates are both imprisoned and promised freedom—rather as the symbolic order is both castrating and grants phallic *jouissance*—spontaneously produces a syntax that could be expressed in the form of a rule: any prisoner who sees that both of the others have black discs can leave straight away, knowing he is a white. Since none of them in fact jump up straight away, each prisoner can be sure that the others do not see two blacks either. In this way, permutation (1) is discounted. Thus, if one prisoner sees that the others have one black and one white disc, he can deduce that he must be a white and make for the door (permutations (1) and (3) having been eliminated). However, if he sees two white discs, he is still dealing with two possible permutations (either (2) or (3)), the only way to distinguish between them being the disc on his own back. And yet, seeing that the other two are not moving should indicate to him that—assuming they are following the same line of reasoning as him—they too do not see one white and one black disc. Once again, he can deduce that he is a white. But what if they all then get up at once? This would cause this prisoner to doubt his reasoning, predicated as it was on the fact that the others *hesitated*. However, if they all then stop again, this second moment of faltering should remove all doubt, for if either of the others could see a black disc on his back, they would not have hesitated for a split second. That is to say, their uncertainty crystallises his certainty. Finally, he can confirm that he must be a white and make for the door with confidence.

What Lacan wants to bring out from this apparently logical solution is the absolute necessity of a subjective dimension of both *hesitancy* and *haste*, and thus of *kairos* as a time of desire that inhabits *chronos* or what might be termed the 'time of the Other'. As he puts it, 'the sophism thus maintains all the constraining rigour of a logical process, on condition that one integrates therein the value of the two *suspensive scansions*'.[32] So, we have a cybernetic machine *plus* the subject caught up in its gears who nonetheless precipitates its unfolding operation. It is already obvious why hesitation is a vital component of this process, but haste is no less important: for if our prisoner were not to act in such a way as to *assert his certainty in a manner that retrospectively justifies it*, he would lose all grounds for any possible certainty. It is this recursive model of causality that will be clarified in relation to the unconscious, repetition and transference in *Seminar XI* almost twenty years later. But already in this 'sophism' of 1945, Lacan teases out the paradox of structural agency:

> What makes this act so remarkable [...] is that it anticipates its own certainty owing to the temporal tension with which it is subjectively charged [...]—so that in the end the conclusion is no longer grounded on anything but completely objectified temporal instances, and the assertion is desubjectified to the utmost.[33]

That is to say, the assertion of certainty, and the act of making for the door, belongs not to a psychological 'I' but to the general logic of the combinatory. This can be seen in the fact that it embodies a 'universal' rule—'one is a white when the others have hesitated twice in leaving'[34]—which applies to all three prisoners even though only one of them instantiates it. If the game of 'odds and evens' demonstrated that 'the very notion of probability and chance presupposes the introduction of a symbol into the real',[35] 'Logical Time and the Assertion of Anticipated Certainty' demonstrates the role of the act in retrospectively constituting a thoroughly conditioned mode of causality, that is nonetheless the only 'real' foundation for subjective freedom.

The Time of Desire and The Time of the Other in Source Code

Turning finally to *Source Code*, Colter Stevens clearly has something in common with the prisoners in Lacan's 'sophism'. He is literally caught in the gears, or circuits, of a cybernetic machine that nonetheless relies on his fragile, fragmented body. Just as the machinic aspect of the symbolic 'parasites' on the *jouissance* of the speaking subject, so Source Code depends on the grey matter of two human brains (Stevens and the dead train-bomb victim, Sean Fentress, into whose memories his neural network is patched). As the film unfolds, moreover, Stevens is increasingly engaged in a competition to secure his freedom from the demands of this programme. He may be a cyborg, but certainly not of the type Donna Haraway has celebrated. The nightmare image of his dismembered body being *incorporated* into the machine resembles instead a hellish update of Marx's early fears about the industrial worker becoming an extension of the machines of mass production. Stevens, then, is also looking for an exit from his cybernetic prison and will only be granted release if he can produce rational data that satisfies the criteria of his keepers: the bomber's name and enough 'intel' to prevent the next attack. Yet from the programme's point of view, Stevens is little more than a string of binary numbers.

But one can see in the film the tension, identified in 'Psychoanalysis and Cybernetics', between the 'universe of scientific precision', which here aims to capture the future before it has happened, and the human time of 'man's waiting' as well as his corollary haste. Yet the intuitive relation between the two is given a twist. For despite appearances, much of *Source Code* is concerned not with haste, but rather with what I would qualify as *urgency*. If haste in 'Logical Time' is '*the moment of concluding the time for comprehending*'[36] in order to pass to the act, what I am calling urgency in *Source Code* is, on the contrary, the imposition of the time of the Other. For this Other, there is no end to the time for comprehending. Whereas haste utilises temporality as the medium by which to enact the act—in other words, as the vehicle for a desire that catalyses change—urgency reduces time to the time of mere *duty*, which is essentially conservative. As opposed to haste, urgency involves the displacement of desire by an unending (repetitive) response to *demand*.

Arguably, *Source Code* could not have emerged pre-9/11, since it explicitly dramatises the strange mixture of permanent exception and risk-aversion that characterises the so-called War on Terror, in which constant vigilance and emergency preparation have become superegoic injunctions. The command centre from which Source Code is operated is called 'Beleaguered Castle'. Urgency, then, would describe something like the dutiful observation of the 'emergency protocols' that increasingly structure everyday life in the beleaguered castle that the West has become post-9/11, ideally without thought or resistance. As a soldier, Stevens is already disciplined for this frictionless enactment of orders. Yet this is pushed to an unbearable limit when his military training is spliced into the binary commands coming from Source Code. This interleaving of military training and responsiveness to machinic commands is obvious from the 'pattern recall' with which Source Code 're-sets' Stevens: he repeats a mnemonic sequence automatically, even in the midst of his profound disorientation, rather as military training is intended to get soldiers through the fear and chaos of battle. The film visually links the use of playing cards in this 're-set' sequence to memories of Afghanistan still lurking in Stevens' cerebral cortex. Yet this mnemonic device is deliberately distanced from subjective memory: all it re-sets him for is the urgency of the Other's agenda, in relation to which his individual past is akin to 'noise' in Shannon and Weaver's sense.

This superegoic dimension of urgency is palpable in the first two-thirds of the film. The train scenario already foregrounds it, since the carriage is full of passengers heading to work, and some are disgruntled that it is running late. Stevens is told that the bomb exploded at precisely 7:48 Central Standard Time. As if to illustrate Freud's link between repetition and the death-instinct, the rendez-vous with this *terminal* destination is made with a monotonous regularity. Each time he is 'sent back in', he re-sets his wristwatch to count down the eight minutes available to him until, once again, he is blown up along with the other passengers. He quizzes one of them who glances anxiously at his own watch: 'You seem concerned about the time. What are you late for?' This is a question he himself would find difficult to answer, since he is *constitutively late*: the bomb itself has always already gone off in the 'reality' that defines the world outside the Source Code programme, where everything he experiences as unfolding in a living present has already happened that morning (as it is put at one point, 'Up here, the clocks only move in one direction').

The bomb is the quintessence of what Lacan would call 'automaton', the repetition that imposes its inevitability with the seeming relentlessness of a machine, often producing the suffering that we call a symptom. It is important to connect this to Stevens' structural guilt, which keeps him locked into this repetition of the attempt to meet the Other's demand. When his will is flagging, he is reminded by Rutledge (the project designer of Source Code and embodiment of the demanding Other) that millions of real American lives depend on his efforts. Indeed, both Rutledge and Goodwin (the operator of the programme who interfaces with Stevens on screen in a kind of inverted Turing Test) repeatedly remind him that attempting to prevent the bomb from exploding is completely 'irrelevant'. It is permanently too late for that. All they want is data that can translate into an arrest, demonstrating thereby Source Code's value as a pre-emptive weapon in the War on Terror (though one, it seems, that requires some initial dead bodies). Pinned uncomfortably to the demand of the Other, Stevens' predicament is captured by Rutledge: 'You are a hand on a clock, understand? We set you, you move forward. We re-set you, you move again. And that represents the entirety of your function here'.

The Other's absolute mastery over his experience of time is indicated early on in his confusion regarding the apparently instantaneous transition from the heat of battle in Afghanistan to what he initially believes to be a military training simulation: Stevens: 'I flew two goddamn sorties yesterday!' Goodwin: 'You've been with us for two months …'. As his abject lack of agency in this situation becomes clearer, Stevens begins to bargain for the only form of subjective time available to him, death (or the completion of it). As a soldier, he is used to viewing death as a potential cost of military service, but precisely as the 'ultimate' sacrifice, i.e. both the most selfless and the last. Stevens responds to Rutledge's clumsy appeal to his sense of military duty by insisting on death as a sacrifice 'ultimate' enough to put an end to this economy of infinite demand: 'Most soldiers I served with would say one death was enough'. As I will argue in a moment, Stevens' motivation changes during the course of the film so that instead of seeking the release of death, he begins to pursue the possibility of a life somehow within Source Code, or within a parallel universe enabled by it (much of the film's ambiguity centres on this unresolved 'or'). And he is right to seek an alternate exit, because unlike the sophism Lacan describes in 'Logical Time', here those keeping Stevens captive are primarily interested in keeping the game running, predicated on him *not* achieving the release of death. Rutledge goes back on his promise to flick the off-switch on Stevens' life-support upon successful completion of his mission. He chooses instead to 'initiate the memory-wipe on SC1 and start crunching the data'.

Not surprisingly, this progressive destruction of agency, and with it the possibility of subjective time, has a devastating impact on the fantasy framework hitherto ordering Stevens' identity. In a posthuman take on Lacan's mirror-stage, this begins early on in the film when, awaking confused on the train for the first time, he beats a retreat to the toilets, only to see in the mirror the face of another (Sean Fentress). Illustrating Lacan's basic point that identity is constituted in an irreducible externality vis-à-vis the bodily experience of the drives, Stevens' ends up pulling an ID card from the wallet he is carrying, which of course shows Fentress's details rather than his own. This loss of identity becomes more serious when the 'capsule' in which Stevens believes himself to be, when not immersed in the virtual scenario on the train, begins to leak hydraulic fluid and shut down its ventilation system, signifying the disintegration of his imaginary co-ordinates. Inside that capsule, he is able to continue to believe that he is a soldier on a simulated mission, even appearing to himself in army fatigues. Yet as the real punctures a hole in that fantasy structure, the capsule itself breaks down until it cracks open, exposing him completely to the unbearable *jouissance* of proximity with the demanding Other. For Lacan, fantasy is a defence against *jouissance* and the absence of the armature it provides can only result in intense anxiety.[37] This is clearest in the example of the psychotic, who often, at least in the case of the schizophrenic, feels invaded at a bodily level by the *jouissance* of the Other because of the foreclosure of the paternal function.[38]

Yet it is probably more accurate to diagnose Colter Stevens (to the extent that one can with a fictional character, which is to say, not all from a properly clinical point of view) not as a psychotic but as an obsessive compulsive. For the obsessive is that type of neurotic that reduces the desire of the Other to a demand to which their very being can answer, and who can therefore conceive of time only as the time of duty in the service of this Other. Like Stevens, the obsessive tends to keep himself in this position of self-mortification with a narrative of guilt. The obsessive is also plagued by automatic repetition, paradigmatically the compulsive washing of hands, often accompanied by a superstitious dread of what will occur if those rituals are not observed. This stems from a deep-seated belief, at once selfless and enormously immodest, that they are responsible for everyone else. Such constructions ultimately enable them to avoid being responsible for their own desire, which they only maintain qua impossible. The obsessive's fundamental question is 'Am I alive or dead?' because he exists only in relation to the Other he serves. In numerous places, Lacan describes Shakespeare's Hamlet as a paradigmatic obsessive. Hamlet is also crippled by the time of the Other (his father's demand for revenge), by a difficulty with putting an end to thinking in order to act (he prevaricates even as Claudius kneels before him at prayer), and by the obsessive's dilemma: 'to be or not to be'. However, if Stevens bears comparison to Shakespeare's creation at the beginning of the film, as the narrative progresses, he tries to find a different exit from the demand of the Other. Strangely, this may be because his obsessive question is answered by Rutledge: learning he is already dead permits him to choose life, albeit in a virtual manner. Thus, Stevens re-discovers the time of his own desire, and thus the capacity to act, by seeking something like a 'cure by love' via the character of Christina.

Paradoxically then, the bomb that represents the quintessence of repetitive inevitability, for which Stevens assumes the guilt, simultaneously forces from him the question of *his own* desire: 'What would you do' he asks twice in the film, 'if you knew you only had a minute left to live?' He has two distinct answers to this question. Initially, he says he would call his father. This relates to his regret about separating on bad terms. One could certainly read this as a rather predictable appeal to American family values, but the source of their disagreement is important: his father did not want him to return to Afghanistan, but Stevens chose duty over love. After somehow managing to phone his father from within the Source Code programme, his second answer sees the reversal of this choice. He now opts for love, in the form of Christina, over duty, in the form of Rutledge and his team of Source Code programmers. Whereas his default answer, as an obsessive, to the question 'what would you do if you knew you only had a minute left to live?' would be 'whatever the Other asked me to do', he learns a different answer from Christina's own response: 'I'd make those seconds count'.

As we have already seen, the issue of what counts is central to modern science and its temporal 'accuracy'. Obviously, Christina does not mean 'count' as the metronomic marking of abstract time, but rather as what possesses value from the point of view of desire. Her answer, which he makes his own, thus pushes Stevens from urgency towards haste. He is able to find room for haste within urgency by exploiting the 'synchronization of watches' at the hidden heart of modern science, which means that the imaginary circulates in the programme and its users. After all, it is Goodwin's humanity that allows Stevens to outplay Rutledge by switching off his life-support machine. In the film's final image–Stevens peering at himself (albeit with the features of Sean Fentress) hand in hand with Christina into the curved reflective surface of Anish Kapoor's Millennium Park public sculpture in Chicago—we seem to have the emergence from the rubble of the disintegrated 'capsule' of a new stabilising fantasy.

This ending is far from unproblematic. I am in no position to judge the validity of the science, but is there not a dubious correlation between super-string theory and the possibility of the 'multiverse' to which the film gnomically appeals, and the mantra of choice in neoliberal capitalism, as if one could simply swap to a parallel universe if one were disgruntled with this one? Certainly, from a Lacanian perspective, it is easiest to see this as a compensatory fantasy papering over the fact of castration, which boils down to there being a flaw in the universe of the speaking being, whichever dimension he finds himself in. There are also ethical questions. Why, beyond the 'common sense' of liberal individualism, should the universe in which Stevens gets the girl be valued above all the others (presumably approaching an infinity) in which he doesn't and everyone on the train dies, as well as everyone effected by the second attack? What, we might also ask, of Sean Fentress, whose mind Stevens has hollowed out to make room for his own, 'parasiting' another's body just as Source Code had done to him?

However, the most crucial problem with the film in the context of the preceding discussion is the way in which it ultimately stops short of its own insights into posthuman temporality and causality. Although it testifies to the symbolic's reliance on the imaginary, *Source Code* ends up implying the imaginary's *supremacy* over the symbolic which once again becomes a prosthetic contingency that can be outrun. Upon experiencing a sense of déjà vu as he approaches Kapoor's sculpture, which he has seen several times before as transient fragmentary images within Source Code, he asks Christina if she believes in fate. She continues to be more insightful than him, when she answers 'Not really. I'm more of a dumb luck kind of gal'. Christina grasps tychic modes of repetition in which the consequences of contingency are subjectivised, whereas Stevens seems reluctant to let go of the reassurances offered by cosmic fatalism at the level of love and the supposed sexual relation. By falling back onto an extremely trite notion of the loving encounter as something written in the stars, even across parallel universes, *Source Code* lapses into fantasy even as it comes close to naming the posthuman symptom. With fate closing like water around the destiny Stevens had made for himself, one would have to agree with Freud's pessimism regarding the 'cure by love',[39] whereby one veils the Other behind an other with whom one believes one fits as comfortably as a hand into a pre-tailored glove. Psychoanalytically informed critical posthumanism must turn this kind of fantasy inside-out, in order to read the logic of the twenty-first-century symptom.

Notes

1. Lacan refers to the *gant retourné* many times, including in *Seminars X, XI* (in relation to Merleau-Ponty), *XIV* (as a refutation of the reciprocal relation between sadism and masochism) and *XXIII* (with regard to Nora Barnacle as a partner-symptom for James Joyce). The original reference actually comes from Immanuel Kant's inaugural lecture, in which he used the example of the inside-out glove to argue that the geometric orientations of 'left' and 'right' have no transcendental relevance when two objects are topologically indistinguishable. While late Lacan is very interested in this same problem in his own use of the planar diagram of the Borromean knot (which is in fact a link, rather than a knot, and which implies two different versions, one 'levorotary' or oriented to the left, another 'dextrorotary' or oriented to the right), he remains consistent in stressing something that interrupts the apparent reciprocity: in *Seminar XXIII* then, it is the button on the glove that means Nora Barnacle can never quite fit Joyce like a glove, even though she is the only woman for him.
2. By 'clinically-informed', I refer to a Lacanian theory conditioned by its psychoanalytic *practice*. Certainly, Lacanian theory exists within academia as a form of cultural theory, but in its merely thematic application, it too often resembles what Lacan himself in *Seminar XVII* calls 'university discourse', which he opposes to the discourse of the analyst. Analysis does involve an interpretation of culture in the form of the social link to which the analysand has an inevitable relation, but such interpretation stems projectively from the knowledge actively produced by the unconscious, rather than from the retrospective application of pre-existing concepts.
3. *Source Code*, directed by Duncan Jones (London: Summit Entertainment, 2011). The film has grossed over $147 million at the time of writing: see http://boxofficemojo.com/movies/?id=sourcecode.htm (accessed July 18, 2017).
4. Hans Moravec, *Robot: Mere Machine to Transcendent Mind* (Oxford: Oxford University Press Inc., 1998).
5. Francis Fukuyama, *Our Posthuman Future: Consequences of the Biotechnology Revolution* (London: Profile Books, 2002).
6. See Ross Benjamin and Von Hilgers, Philipp, *War Games: A History of War on Paper* (London: MIT Press, 2012); Andrew Pickering,

The Cybernetic Brain: Sketches of Another Future (London: University of Chicago Press, 2011); John D. Steinbruner, *The Cybernetic Theory of Decision: New Dimensions of Political Analysis* (Oxford: Princeton University Press, 2012); and Philip Mirowski, *Machine Dreams: Economics Becomes a Cyborg Science* (Cambridge: Cambridge University Press, 2001).
7. Lydia H. Lui, *The Freudian Robot: Digital Media and the Future of the Unconscious* (London: University of Chicago Press, 2010).
8. Jacques Lacan, *Seminar II, The Ego in Freud's Theory and in the Technique of Psychoanalysis*, trans. Sylvana Tomaselli (London: Norton, 1991).
9. Jacques Lacan, *Seminar XI, The Four Fundamental Concepts of Psychoanalysis*, trans. Alan Sheridan (London: Norton, 1998).
10. Jacques Lacan, *Écrits*, trans. Bruce Fink (London: Norton, 2006), 161–175.
11. N. Katherine Hayles, *How We Became Posthuman: Virtual Bodies in Cybernetics, Literature and Informatics* (London: University of Chicago Press, 1999).
12. Liu, *The Freudian Robot*, 165.
13. Three or so years after the Macy Conferences, Lacan says explicitly, 'Read anything by Mr. Norbert Wiener; its implications are huge': see Lacan, Jacques, *Seminar III, The Psychoses*, trans. Russel Grigg (London: Norton, 1997), 37.
14. Danielle Eleb, *Figures du destin: Aristote, Freud et Lacan ou la rencontre du reel* (Saint-Agne: Erès, 2004).
15. Jacques Lacan, *Le Séminaire Livre X, L'angoisse* (Paris: Éditions du Seuil, 2004), 180. An English translation of this seminar is now available: see Jacques Lacan, *The Seminar of Jacques Lacan: Book X, Anxiety*, trans. A. R. Price (London: Polity Press, 2015).
16. Muller and Richardson (eds.), *The Purloined Poe: Lacan, Derrida and Psychoanalytic Reading* (Baltimore: Johns Hopkins University Press, 1988).
17. Jacques Derrida, *The Post Card: From Socrates to Freud and Beyond* (London: University of Chicago Press, 1987).
18. Lacan, *Seminar II*, 205.
19. When Lacan speaks of the 'materiality' of the signifier, he means something quite distinct from the recent theoretical innovations that can be collected under the heading of the 'new materialism'. Broadly inspired

by Alain Badiou and his student Quentin Meillasoux, but evolving beyond their work through the Speculative Realism movement associated with Graham Harman, these new materialisms are explicitly ontological: they relate to being *qua* being, supposedly beyond the epistemological constraints imposed by Kantian critical philosophy. In the form of object-oriented ontology in particular, 'materialism' emphasises the productive relations among non-human objects quite independent of the knowing human subject. For these reasons, the 'new materialism' feeds into posthuman themes of the technological production of subjectivity and the distribution of consciousness via a (digitally) 'extended mind', and so on. Lacan, however, makes it very clear in *Seminar XI* that the psychoanalytic hypothesis *par excellence*, the unconscious, implies not an ontology but an ethics. As such, the unconscious, as instantiation of the 'materiality of the signifier', is a rupture with (imaginary) being, rather than a 'tipping point' or 'quantum leap' emerging from a particular arrangement of being conceived as matter. Where the 'object' in object-related ontology is a 'thing' embedded in a co-constitute web of productive inter-relations, the 'object' for Lacan is always an effect of the symbolic structure inhabited by the *speaking* being.
20. Lacan, *Seminar II*, 192.
21. In *The Interpretation of Dreams*, Freud argued that it was impossible to free associate random numbers because significance could always be retrospectively ascribed to the choice. See Sigmund Freud, *The Interpretation of Dreams: The Penguin Freud Library*, Volume 4, trans. James Strachey (London: Penguin, 1991), 659.
22. Lacan, *Écrits*, 731.
23. Lacan, *Seminar II*, 296–308.
24. Ibid., 297.
25. Ibid., 298.
26. Ibid.
27. Ibid., 300.
28. Ibid., 306.
29. Jacques-Alain Miller has described this as science's tendency to 'eat the real'. See Jacques-Alain Miller, 'A Fantasy', *Lacanian Praxis: International Quarterly of Applied Psychoanalysis* 1, (May 2005), 5–16.
30. Lacan, *Seminar II*, 307–308.
31. Lacan, *Écrits*, 163.
32. Ibid., 165.

33. Ibid., 171.
34. Ibid., 172.
35. Lacan, *Seminar II*, 182.
36. Lacan, *Écrits*, 169.
37. Lacan, *Séminiare X*, 55.
38. Darian Leader, *What is Madness?* (London: Hamish Hamilton, 2011).
39. See Sigmund Freud, 'On Narcissism' in *On Metapsychology: The Theory of Psychoanalysis, The Penguin Freud Library*, Volume 11, trans. James Strachey (London: Penguin, 1991), 59–97.

6

A Fly in the Appointment: Posthuman-Insectoid-Cyberfeminist-Materiality

Ben Woodard

Introduction

Theorists such as Jussi Parikka, Jakob von Uexküll, Eugene Thacker, Sadie Plant, and others have utilized the figure of the insect as a particularly salient way of reading the materialization of information. Certain affinities of insect anatomy and behavior (such as antennae, segmentality, and swarm behavior) with technology aesthetically collude with technologies of communication (nanorobotics, military drones, surveillance devices). But is such use of the insect figure merely metaphorical, or does it drag with it other (less functional, less descriptively desirable) aspects of the insect body and its bearing on the physicality of information? In either case, the notion of the insect body (as the actually biological entities that also carry aesthetic and cybernetic valences) proves an interesting point for recognizing a tension between communication

B. Woodard (✉)
Institute for Philosophy and Art Theory, Leuphana University, Lüneburg, Germany

S. Matviyenko and J. Roof (eds.), *Lacan and the Posthuman*, The Palgrave Lacan Series, https://doi.org/10.1007/978-3-319-76327-9_6

(as information) and language (as speech) in contemporary posthuman discourse. As Lacanian psychoanalysis is invested in this difference, and in unfolding the post-, in-, or non-human within the human, a guiding question issued to Lacan in the name of the insect body could be: Did he have a vested interest in the non-discursive or "primordial" materiality of the world other than as an instructive disruption of language? In other words, does materiality (insectoid or otherwise) have a role in his theories other than that of the speaking of/as the symptom, thereby cutting through, but ultimately mutating, communication?

The link between the materiality of the insect body and Lacanian psychoanalysis can be illustrated through an investigation of various affective engagements with insects and technology, such as the numerous bug-human-machine hybrids in the films of David Cronenberg. While beginning from a mainly critical stance on Lacan, this essay mines the potentiality of a posthuman engagement with insects as it appears, however negatively construed, in Lacanian psychoanalysis. This imbrication indexes the relation between insects and language as well as between insects and cybernetics (connections Lacan himself draws) and then builds off of a critique of Lacan and of the posthuman use of insects in Rosi Braidotti's and Gilles Deleuze's work via a detour through cyberfeminism, focusing particularly on the work of Sadie Plant. But why focus particularly on the insect's relation to psychoanalysis to address the posthuman?

While Lacanian psychoanalysis purports to be a non-humanism, it does not obviously nor immediately provide an adequate engagement with materiality and the materiality of information to address an anti-humanism or anti-anthrocentrism that is technologically, or otherwise futurally, realizable. Yet psychoanalysis appears posthumanist by resisting the conservatism of modern philosophy as well as the futurist exuberance of much posthumanism (in that "we were never rational," and we cannot surpass an irrationality, or split subject or Kantian x—which we do not, and most likely cannot, understand and so on). This not purely formal articulation of the subject (as excess or as fundamentally indefinable) suggests an informatics already swarming over the world, a material code working as a necessary groundwork for a posthumanism that attempts to be more than metaphorical window dressing.

These tensions, however, tend to get bogged down in determining who is *more or less* posthuman versus posthumanist. Both N. Katherine Hayles and Cary Wolfe accuse the other of maintaining a humanism when discussing posthumanism. Hayles' *How We Became Posthuman*, admittedly sarcastic in its title, critiques certain aspects of posthumanism, arguing primarily that the trans- and posthumanist thrust to leave the body behind is a covert liberal-humanist move, which reifies the false universalism that it purports to abandon.[1] Her assertion that posthumanism can be reduced to subject construction is questionable, as is what Hayles proscribes to combat this form of the subject—that is, a dialectical relation of information to embodiment.[2] In his introduction to *What is Posthumanism?*, Wolfe takes Hayles to task not only for externalizing processes such as mutation, but for setting up a dialectical antithesis between information and matter.[3] He claims that, while Hayles critiques posthumanism for its opposition of bodies and embodiment, she does not seem to recognize any sort of serious ontological connection between the materialized and the unmaterialized. The case of mutation, as Wolfe notes, appears as a kind of interruption of a pattern, yet, it would seem difficult to discuss patterns of being that do not have an ontological status where, instead of mutation being evental, mutation is merely an effect of the various "groundings" of patterns in material things. For Wolfe, "bad posthumanism" is a humanism which is rampantly transformative of the material conditions of actual human beings. "Bad posthumanism" seems inseparable from transhumanism in that both purportedly assert that technological progress will allow us to transcend the limits of our biologies. Hayles, on the other hand, critiques this very same kind of posthumanism for remaining too humanist in the sense of privileging the disembodied universalist subjectivity of the humanist project. But Hayles' suggestion of pattern over information does not solve the problem nor, would I argue, does Wolfe's utilization of Jacques Derrida's trace as against the supposed digitality of technological inscription.[4] While both attempt to resist the sweeping power of technology in discussions of the posthuman, this resistance often seems to take place in terms of being a resistance *in principle* rather than in a more material sense. For both, gradual changes due to technological prosthetization as well as developments in cognitive science seem to slip under the radar. This is possible because much of posthuman theory relies too heavily on a discursive, or human-produced, concept of materiality instead of a materiality preceding and composing human entities, if affected by human actions. In other words, the posthuman problem orbits around an axis of materialism and materiality where the stricter definitions of the latter are mixed indiscriminately with the Deleuzian and Marxist meanings of the former. The broadly Marxist view tends to limit the "active" take on matter common in the Deleuzian tradition and taken up and expanded by new materialist thinkers such as Jane Bennett, Karen Barad, Manuel DeLanda, and many others. In these related traditions, the measure of activity, one might say the behavior of materiality, is not determined by human subjects but forms subjectivities without their complete knowledge. While Marxist theory emphasizes the importance of an epistemological relation to materiality (at the very least in terms of the conditions of labor or of capitalist survival), new materialist theory, at least in spirit, sets up a materiality that determines knowledge and limits activity before we could say to know it or manipulate it.

From this distinction, one can see why the insect is such a fascinating figure caught between these formulations of the material and how it lights the way for an engagement with Lacanian psychoanalysis. Lacan himself invites the connection between insects and knowledge with his brief mentions of swarms and bees, and as Jussi Parrika has shown in *Insect Media*, the insect body unveils a strange archeology of the production of media, as the materialization of communication but perhaps not language, in the late nineteenth and early twentieth centuries.

Lacan's Bees (Bodies and Images/Letters and Codes)

Circumnavigating the contemporary influence of Slavoj Žižek, it becomes difficult to square the philosophical, or at least more vaguely theoretical label, of materialism with Jacques Lacan. Lacan's oeuvre is dominated by mathematical and linguistic games and is replete with artistic, literary, and playful turns of phrase, all of which mark the overwhelmingly clinical, and at least functionally human, nature of Lacanian psychoanalysis—materiality is a question of the form of the subject. The tension around materiality in Lacan's *Écrits* is most evident in the relation of science to psychoanalysis.[5] This takes place in the discourse of the subject, or of the subject of science in relation (or non-relation) to the subject who performs science (a subject that seems rankly Cartesian). In "On the Subject Who is Finally in Question," Lacan states: "It is obvious that psychoanalysis was born from science. It is inconceivable that it could've been from another field."[6] While Lacan could simply be referencing the nascent neurological work of Freud (who, contrary to the claims of his critics, was a skilled practitioner of the cognitive sciences, diagramming nerve cells, for instance), it seems Lacan is trying to absorb the efficacy of scientific practice while moving beyond its subject (as a supposedly naïve limit). Lacan constructs a posthumanism by robbing the subject of power over its own mind as well as the productions spewing forth from that mind. However, Lacan is able to do this only via a subject formalism and not without mythologizing or appealing to the occult (against his own wishes) regarding the relation of thought (and psychoanalytic practice) to the real. Žižek, among others, is at pains to defend Lacan as an anti-relativist, stating that Lacan dismisses the scientific real as namely that which we lack.[7] The weirdness of insect materiality amps up the stubbornness of the flesh against the prophylactic of the lack, while also downgrading the privilege of language over communication central to many streams of posthumanism. But certain Lacanian attacks upon posthumanism perhaps overplay the anti-scientific reading of Lacan's thought. Ted Hiebert in "The Lacanian Conspiracy" takes issue with Hayles' (and, by extension I would argue, Wolfe's) above-mentioned rhetoric of resistance—where posthuman texts merely display the stubbornness of the human. Hiebert argues that the slogan of the posthuman should be "placebo ergo sum" (I hallucinate myself into being).[8] Yet, the developments of analytic thought and science as well as continental thought (posthumanism and otherwise) have already proposed more radical challenges. For Thomas Metzinger, the very form of self-perception is illusory—one has entered the illusion before even being able of constructing the illusion.[9] In his work, Metzinger furthers Choang-tsu's worry about whether the self is dreaming it is a butterfly or whether it is a butterfly dreaming it is human. Hiebert moves around Descartes' dismissal of the trickster demon which inhabits the same skeptical space as the dreaming butterfly (as well as the brain in the vat). In these cases, there is a non-choice before one can even start to think of who or what is dreaming who or what. Below, we will see that while Gilles Deleuze appears to come close to such a position at times (at least as his nomadic subjectivity is articulated by Rosi Braidotti), there still remains an unnecessary ghost of a rational subject. The central problem of the ghost lies in the dismissal of a generative materiality in both Lacan and Deleuze, albeit in very different ways.

Lacan's lack of interest in active materiality makes his posthumanism halfhearted or, at best, a posthumanism which simply exists to caution against itself, to set posthumanism apart from transhumanism or the horrific possibilities of being posthuman (in the flesh). In his text *Subjectivity and Otherness*, Lorenzo Chiesa lays out the following schema for how materiality is dealt with by Lacan. Chiesa writes: "The Real that psychoanalysis deals with is the Real-of-the-Symbolic, which is not to be confused with everyday reality and should also be clearly differentiated from what Lacan calls 'the primitive Stoff,' matter unmediated by the symbolic."[10] Lacan connects *stoff* to a metaphor of a hydroelectric plant as the unconscious—the energy of the river is meaningless for Lacan until it reaches symbolization.[11] Chiesa's reading of Lacan's use of the term *stoff* would suggest that Lacan's view of materiality ignores the turbulent and eruptive nature of materiality in, and as, information—particularly as it shows up in the insectoid body.[12] The kind of materiality Lacan appears to be interested in is materiality in terms of *material conditions* of human existence, in a largely Marxist sense, though perhaps including more than only economic constraint. While arguably more evident in Sigmund Freud's texts than in Lacan's, the question of trauma ("What happened?") is a question which concerns the subject's perceptions and attachments feeding into, and resulting from, that perception, but one that is interested in the social, political, and even directly physical constraints of experience.

Following Jacques-Alain Miller, however, we could further introduce a distinction between the materiality of the world and the materiality of the lived body, not in a phenomenological sense but in a biological one. In his "Lacanian Biology and the Event of the Body," Miller suggests that the stubbornness of the concept of life under analysis merges instructively with Lacanian psychoanalysis.[13] One aspect of this merger is the very problem of merging, of the idea of life as individualized bodies and as a continuity of drives and forces creating and moving through those bodies. Discussing Diderot's text *d'Alembert's Dream* Miller writes: "[the dream] begins with the image of a swarm of bees described as a clump that appears as a being, an individual, an animal. It is evidently an illusion. It is an assemblage, but, if we blur the little legs the bees are holding on with, if we pass insensibly from contiguity to continuity, we can see a whole and an animal-one."[14] It becomes difficult to separate the human-centered view of the world, even at the generic level of trauma and/or jouissance as caused by material incursion, from the formations of various materials into creatures or entities. Miller points out that this is in part an injunction against Merleau-Ponty and Heidegger who, he claims, are attempting to restabilize soul and body, the human and nature.[15] But to count Merleau-Ponty and Heidegger as the same in the instance of the body seems to be an error, a shortcut to get back to the human as fundamentally disjointed from nature at the level of the material, but again, a material that is about human conditions already cut from nature. Since the moment of the formation of the individual is equated with the recognition of the image in the mirror stage, it is interesting that so many non-human entities such as locusts swarm cross Lacan's essay (along with pigeons). For Lacan, the animal (and in numerous cases the insectoid) marks the organism relating to reality self-reflexively via its own image in nature, whereas Lacan grants only second-order reflexivity to the human. Referring to this first order of the *imago*, Lacan writes:

> [I]n the case of the migratory locust, the shift within a family line from the solitary to the gregarious form can be brought about by exposing an individual, at a certain stage of its development, to the exclusively visual action of an image akin to its own, provided the movements of this image sufficiently resemble those characteristics of its species.[16]

As Deleuze and numerous followers of his would be quick to point out, Lacan ignores the swarm-like being of the locusts; the very way the locusts behave has usurped the fixation of the signifier attributed to Lacan. Or pointing back to d'Alembert's dream, one cannot simply liken the assemblage to a clump of continuity.

In her text *Metamorphoses*, Rosi Braidotti explores the becoming-insect aspect of Deleuze's work. Braidotti discusses industrial ants, destructive crickets, and bees as real-life factories.[17] While she discusses David Cronenberg's *The Fly*[18] and addresses broad connections between insects, women, and technology,[19] Braidotti focuses on the sonic weirdness of insects on the one hand,[20] and on the other, on queer sexualities which fellow Deleuzian feminist Elizabeth Grosz finds in them.[21] Ostensibly, Braidotti's anti-psychoanalytic thrust following insects boils down to the following: While they suggest radical forms of becoming, in Gilles Deleuze and Felix Guattari's work, this becoming is not adequately accounted for given the sedimented contours of male subjectivity.[22] The problem is then why and how Braidotti relies on a form of subjectivity that seems less realist than even a severely tweaked form of Deleuzian–Darwinian biophilosophy,[23] but is more than Deleuze's affective subject.[24] In other words, Braidotti seems to solidify or physicalize subjectivity more than Deleuze in order to get at the real split of gender yet this is incompatible with the genericity of Deleuzian subjectivity in its more panpsychist moments. Braidotti's critique seems to stay at the level of Lacan in terms of the substantiality of the subject as a form, and in so doing, she pushes aside the productive materiality of what makes the illusion of the subject possible at all. There are two ways of defending Lacan from critiques such as Braidotti's: One is arguing that he in fact takes the developments of science more seriously than would appear; Adrian Johnston will be taken as defending this view. The other is that Lacan in fact does have a place for affect; Joan Copjec is taken up as the defender in this case.

In his excellent essay, "The Weakness of Nature," Adrian Johnston points out the tendency of Lacanian-inspired theorists to overemphasize psychoanalysis as a practice in relation to the signified and how this relates to the two images of man in Freud. The essay differentiates between the "bad" Freud, who desired a kind of biologically inspired analytic of neurological processes, and the "good" Freud, who discussed how the representation of man disrupted this purportedly natural image.[25] Johnston proposes that a Lacanian materialism or naturalism should be taken as a kind of Trojan horse into the core of the sciences. This could serve to effectively denature nature itself.[26] Johnston's essay speaks to the titular weakness of nature. It follows from his claim that since nature breaks down (in traumas, anxieties, and so on), it is not seamless but instead fractured and contradictory. In the process of describing nature thusly, Johnston makes a move common to psychoanalysis. He equates nature (writ large in a cosmological sense) with biologically determined human nature. Johnston, along with many others, takes the purportedly naïve attitude of the former and inflates it in the latter, thereby covering over the very equation of non-human nature and human nature, or covering over cosmological nature with the human grasp of what is natural for us. This shell game is malfeasant in that it then takes what could be easily characteristic of flawed human behavior (such as the pleasure principle) and writes them back into a nature in order to make it flawed fundamentally. In so doing, the teeth of psychoanalytic analysis are sharpened by, while simultaneously made of, a nature that is only productive of the subject insofar as that production has nothing to say about its capabilities. Despite Johnston's defense and interesting discussion of Lacan's metapsychological theories and their relation to recent developments in neurobiology, it becomes difficult not to see Lacan as trying to have his cake and eat it too, so to speak. A rupture in these synthesized images appears in particular in Lacan's frequent reliance upon the biological sciences. In discussing the relation of language to code (and to return to insects), Lacan makes a well-noted remark on bees. He writes:

> It is now generally recognized that, when a bee returns to its hive after gathering nectar, it transmits an indication of the existence of nectar near or far away from the hive to its companions by two sorts of dances. The second is the most remarkable, for the plane in which the bee traces out a figure-eight—a shape that gave it the name 'wagging dance'—and the frequency of the figures executed within a given time, designate, on the one hand, the exact direction to be followed, determined in relation to

> the sun's inclination (by sensitivity to polarized light), and, on the other hand, the distance at which the nectar is to be found up to several miles away. The other bees respond to this message by immediately setting off for the place thus designated.[27]

Lacan states that language is more than a symbolic representation (more than a metaphor), thereby questioning whether the bee's dance can be considered a language. Lacan uses the quintessential example from speech act theory, that of a marriage vow, to demonstrate the difference between language and communication (or speech). He argues that speech is non-functional, whereas language is purely functional/relational.[28] Lacan is unabashedly anthrocentric in his assertion that language is a privilege of the human.[29] He distances speech from language in saying that while the former informs, the latter invokes.[30] The line drawn between invocation and information (between perhaps what we could call affect and effect) becomes further troubled in Lacan's movement across various subjects of, and subjects doing, science (which may invoke a cybernetic relation to materiality as such). Lacan moves from the subject of science to the subject of the "I" in "On the Subject Who Is Finally in Question" and "The Function and Field of Speech and Language in Psychoanalysis."[31] He then selectively synthesizes the two in "Science and Truth." Lacan writes: There is no such thing as a science of man because science's man does not exist, only its subject does. My lifelong repugnance for the appellation "human sciences" is well known; it strikes me as the very call of servitude. But the fact is that the term is also incorrect, except in the case of psychology which has discovered ways to outlive itself by providing services to the technocracy.[32] To return to "The Function and Field of Speech and Language in Psychoanalysis," Lacan notes the incapacity of the cybernetic to grasp language.[33] Specifically, Lacan notes that no cybernetic device is capable of turning a response into a reaction.[34] This brings us to the affect defense put forth by Joan Copjec. In her essay, "May '68, The Emotional Month," Copjec relates the place of affect in Lacan's works to his critique of a kind of cybernetic fantasy.[35] Copjec argues that despite Lacan's apparent structuralism (and, strangely, because of it), affect has an important role in psychoanalytic discourse. Affect animates things;

it is the surplus that is movement as psychic discharge.[36] Copjec argues that Lacan's technologically enhanced cyborg version of the lamella is constructed to show that even a human entirely plugged into the exterior can still be disrupted by pathologies, jouissance, and other mental disruptions.[37] This interface fantasy, as André Nusselder puts it, is externalized in the early days of cybernetics (in the belief that the exterior world can be purely informatically captured) and then reversed in the ongoing cybernetic revolution (in that technology can allow us to escape ourselves into an even greater outdoors, that of the net).[38] In many ways, Copjec's essay is a diagonal Deleuzian defense of Lacan. Deleuzian scholars are showing the affective affinity between the two. This is particularly evident in the somewhat obscure status of affect on the one hand and the privileging of particular affects, such as shame and anxiety, on the other.[39] The flipside is to overcode code itself with meaning or intent always already humanized, which Deleuze can be guilty of, however much he first appears to be arguing the opposite. Whereas Lacan leans on evocation to signal the more-than-ness of language (over speech), Braidotti (following Deleuze) utilizes expression: "Expression is about the non-linguistically coded affirmation of an affectivity whose degree, speed, extension, and intensity can only be measured materially, pragmatically, case by case."[40] This appeal to the actual of the individual serves as a counterweight to the Deleuzian emphasis on the virtual as well the ambiguity of both the virtual's relation to its instantiation and the ambiguity of affect's relation to both the actual and the virtual. To better address the problem of informatic instantiation, I will examine David Cronenberg's film from 1986, *The Fly*.

Cronenberg's Burroughsian Bugs

In his piece "Affects," Stuart Schneiderman playfully notes that affect is the buzzing of the fly around a psychoanalyst as they attempt to write about clinical practice. While affect is an annoyance (in its excess) to the strictures of psychoanalysis, Schneiderman's note points to the fly (and the insect more broadly) as an interrupting informational speck of materiality.[41] In David Cronenberg's *The Fly*, a scientist (Brundle) attempts to teleport himself with his impressive invention. Through a malfunction, he creates the odd amalgam of Brundle-Fly, a technological glitch—or one may say bug—which causes the weird or non-embodiment of a literal bug (the fly). The telepods (the devices that are designed to move an object from one place to another) work from the premise that the matter of any living thing can be taken apart and then be put back together in another place.

Cronenberg's film has been taken up as a popular text in posthuman texts due to its emphasis on gender, transformation, and identity caught between technology and its technological reduction. While Jussi Parrika dismisses *The Fly* as too easy a choice to think the posthuman,[42] the film seems interesting enough to warrant attention. It is not a matter of embodiment versus informatics or of shifting that division to materiality versus immateriality (as Luciana Parisi argues[43]), but in addressing the informational realm itself, the advent of Brundle-Fly as a categorical error functions on a different level from his material being. While one could say that a genetic mishmash causes Brundle to become Brundle-Fly, this is not to say that it is the shifting of information as the shuffling of the pages of a manuscript. In a passing reference to Cronenberg's *The Fly*, Hayles states that mutation "marks a rupture of pattern so extreme that the expectations of continuous replication can no longer be sustained."[44] Pointing out that mutation functions similarly to castration, she feeds the torqued genesis of the film's scientist back into a discussion of the symbolic and away from the genesis of forms and the formed. This is gender troubled by Jerry Flieger as the image of the newly emerged fly speaks to the viscera of abortion.[45] *The Fly* is also deserving of attention here in relation to psychoanalysis, because it contains a Lacanian mirror scene in which Brundle-Fly sets asides his unused organs (a Deleuzian moment for sure) and realizes that he has *always been* a fly. *The Fly* suggests the informatic nature of materiality, subsequently referencing an ontology which suggests real patterns, not patterns or ideas as simulacra (in a naïve Platonic sense) but ideas as the patterning units of material becoming itself. The film portrays the genetic as instructional software which is instantiated through the body (mediated by the telepods) as contextual hardware. By holding information and materiality apart as a critique, however, Hayle seems unwilling to bring them back together in any significant way. The film likewise maintains an odd separation as well as Brundle, before becoming Brundle-Fly, has to make certain adjustments to account for flesh (as opposed to inanimate things), thereby making flesh a more complex entity, a complexity justified in a vaguely humanist sense. This may be a result of Cronenberg's long-time fascination with the flesh (or replacement of actual flesh with a plastic cinematic flesh) as well as the horror of the transmission of pattern. Brundle becomes Brundle-Fly because of the telepod computer attempting to reconcile two incompatible forms of information at the genetic level.

The fly as contaminant and as transmission appears in numerous horror films such as in *The Ring*. When Naomi Watts' character inspects a film which, unknown to her, transmits the suffering of a long dead girl, she sees a fly on the tape that she picks up off the screen plucking it from the image into reality. Beyond this, the fly has a weird history as a kind of attendant of base materiality (perhaps feeding Bataille's spider of the formless), as indexing the sort of grossness of life but also the emergence of particularity from grossness. We could say that there is the fly of the monster (of contamination), the fly of emergence (of the invisible maggot), and the attendant fly of the formless which is the creaturely fly—creaturely as neither monster nor animal, as discussed by Eugene Thacker.[46] We could say that the first fly is the disrupter of known patterns (at the level of thought), the second being the disrupter of materiality (it seems to come from nowhere, as abomination), where the third indexes both and suggests that there are patterns at the ontological level which can give form, depending on the strictures of various physical constraints—life which can be without us or beyond us in Thacker's articulation. Thierry Bardini has argued in his text *Junkware* that the capacity of information transmission in genetics overrides the concern of the actual materiality of genetic transfer; the transmission capacity of materiality dissolves it. To circle back to the insectoid, the central amber-trapped mosquito of *Jurassic Park* (carried about in the walking stick of the park's primal father John Hammond) represents the pure potentiality of DNA. By extracting dinosaur DNA from the mosquito, Hammond's company InGen is able to resurrect the long dead creatures. However, this requires supplementation from creatures of the modern world (namely frogs) whose ability to spontaneously change their sex leads to unregulated procreation in the wild in which Sam Neil's character, Dr. Alan Grant, providentially notes that "nature found a way" echoing the warning of Jeff Goldbloom as a chaos theoretician. The disregard for the embodiment of information returns with a vengeance in the emergence of female dinosaurs. As Luce Irigaray has analyzed in her response to Plato's allegory of the cave, embodiment has historically been devalued through its feminine association. The purported transmateriality of information relies upon, yet at the same time seems to often forget, the feedback loops (both positive and negative) of the environment as well as the parts of the organisms itself as ecology. As Manuel DeLanda has made in an oft-cited point, material shifts in land masses, for instance, can determine the genetic future of a species or set of species in a larger ecosystem.[47] To move from *Jurassic Park* to Jussi Parikka, the latter supplants the transmateriality of the fantasy of the former with an archeology in which he complicates the relation between ontology and epistemology. One difficulty here is the metaphorical baggage carried by describing technology as insectoid[48] while attempting to maintain some form of materialism. This is particularly difficult in the realm of the life, where existence is equated with excess[49] in which life becomes (purportedly) without structure.[50] The aforementioned trinity of flies (lines of "flyght") speaks to immanence, sense, and extilligence or the formless, the monstrous, and the abominable emergent. What is missing is the realm of thinking and being seemingly closed off from thinking and any recognizable form of being, that of the real or life in itself, that unheard and energetic capacity often claimed by vitalism, Deleuzian or otherwise. This cluster of kinds of patterns and grounds operative in life, highlighted by weird forms, is indicative of Deleuze and Guattari's nomination of becoming as such as becoming animal. Becoming (fly) always involves a multiplicity, and as Deleuze and Guattari write, "Beelzebub is the devil, but the lord of the flies."[51] But Deleuze and Guattari operate in the regime of sense and are unwilling to cast too speculative of an anchor into that bubbling morass of immanence (or the given without givenness) of the sick formless matter of creaturely life. Gradations too quickly become affective intensities when, in the metaphysical pot, they are more accurately onto-epistemological distinctions. These gradations are not simply a virtual intensity defined as not-matter or not-information but the capacity to be both, a capacity that determines human choice and is not determined by human action. Any sense of becoming-insect must reject a lamination of the subject (or of sense) which would shield it from the churning weirdness of both pattern and information (but an immanence given without givenness) but, at the same time, is cut off from thought through the onto-epistemological obscurity of its being anterior and non-sensible but only thinkable in the speculative or metaphysical fashion. This thinkability is possible due to the fact that the subsequent groundings and ungroundings of the force of the real (while inexact) capture and repeat some pattern which functions in the real itself.

Beyond the limits of such surging becoming there is the real ground which has constricted it and the metaphysical ground which obscures those bounds and the possibility of another ground or grounds. The psychoanalytic operation is the anthropomorphic annihilation of these grounds in service of the extilligent, whether linguistic, juridical or otherwise *human*, informatics (or at least a drastic tactical minimization of these grounds according to Chiesa).

The creaturely, on the other hand, indexes the theoretically decisional nature of thinking, and thinking life. To return to Brundle, and Brundle-Fly, the informational reduction of man, fly, and man-fly, is only a small step away from cloning, despite the freeform style of Cronenberg's science-fictive presentation of genetic tinkering. Brundel's telepod attempts to circumvent Hayles' dialectic of body and embodiment through the informational within itself—with the initial failures of the experiments being the result of some incalculability of the flesh. But there is nothing, to stick to the absolute as series of materializations following from Platonic ideas-as-attractors, to make the flesh different in this sense (at the level to confuse the telepod hardware) but only if this flesh is all together a different kind of meat.

Cronenberg's cinematic imaginary is more akin to Lyotard's great ephemeral skin as a libidinal materialist materiality, a flesh already maggoty with the subterranean immanence given without givenness. His 1983 film *Videodrome* goes beyond *The Fly* as it enacts a libidinal explosion of Kantianism—with both being the tumor-inducing externalization of illusions (transcendental illusion). Against Lacan, and at the level of the species as evolutionary, materiality is glazed with an energy that seems specific to it, to the intensities of that particular psychology which has resulted from both material conditions (in the non-libidinal sense, in the real immanence yet untouched by human thought) and libidinal materiality (the psychological feedback of thought on itself). The posthuman is an outfolding of the reeking mass of life; it is the limit of, yet also fuel for, thought. It is, in the end, perhaps unsurprising that Aquinas states: "All the efforts of the human mind cannot exhaust the essence of a single fly."

Difference Engines or Machine Women to Cybernetic Insect Women

In *Metamorphoses*, Rosi Braidotti sets out to investigate the "women-insects nexus"[52] but does not seem to get much further than suggesting a certain queerness, both sexually and sonically, coming from our knowledge of insects. Sadie Plant, founder and once member of the Cybernetic Culture Research Unit (the prehistorical reservoir of some strains of speculative realism), more directly addresses this nexus through the history of computing, particularly in her text *Zeroes and Ones*. The insect and the informational are tied together aesthetically and at numerous deeper or more functional levels than Bradotti's analysis suggests. As Plant has pointed out, it was Grace Hopper, who worked on the machine language COBOL during the 1940s, who coined the term "bug" for a computer glitch, having discovered an actual moth in the machine.

Plant's fascinating text *Zeroes and Ones* makes it clear that the widespread exclusion of women from computer science requires first and foremost a retroactive historical erasure. The text begins with Ada Lovelace, the enchantress of numbers, who effectively wrote the first computer program for Charles Babbage's unfinished analytic engine, and eventually moves on to figures such as the aforementioned Grace Hopper. The central issue of Plant's book for the task at hand is women as the embodiment of microprocessing or micropower in place of, then in tandem with, and then within, computers. As Plant outlines, the practice of weaving (which led, albeit indirectly, to the first use and systematization of punch cards at the dawning of the nineteenth century) illustrated women's vested interest in the machinations of reproducibility and, beyond this, the willingness to believe in the analytic powers of machines, powers which thereby threatened the intellectual as well as physical work of men.

In the early days of computing, whether it was instances such as Lovelace, Hopper, or the ENIAC team,[53] women worked within machines. One can easily think of a Vertov-like dream where scores of women worked the telephone switchboards (and the telegraph before this) literally embodying that technology (to say nothing of the typewriter, word processor, or stenograph). Women as computers held the place which would eventually become the issue of the ghost in the machine, a haunting which erases the computer's former inhabitants. Plant's position here can be problematic. As Zoe Sofoulis puts it in her essay "Gender Quake," she can appear as a strange mix of hyper-anti-essentialist Haraway and essentialist Irigaray.[54] Yet this is not simply an error on Plant's part but a serious consideration of the negative aspects of a Deleuzian inspired politics and ontology (and their tendentious relationship). In her essay "Binary Sexes, Binary Codes," Plant discusses the relation between two senders in a communicative network as often being gendered and argues, following Irigaray, that women have often been coded as that which is sent—the exchange of women being the basis of familial bonds in exogamic relationships.[55] The economics of materiality in exogamy question the reliance on the extimate as well as ek-sistenz in Heidegger, as Braidotti argues[56] (here it is hard not to think of Cronenberg's *Existenz* as a corporealization of the exteriority of fantasy). For Deleuze, again according to Braidotti, affectivity itself structures subjectivity.[57] But this ignores the materialist basis for Plant's equation of patriarchy with economy.[58] In "On the Matrix: Cyberfeminist Simulations," Plant writes: "Cyberfeminism is an insurrection on the part of the goods and materials of the patriarchical world."[59]

In *Zeroes and Ones*, Plant discusses Irigaray's claim that woman is pure mimicry[60] reflecting her earlier comments on Anna Freud at the loom[61] and the "problem" of feminine masquerade—namely that the creation of women's clothes is a copy of the matted pubic hair of nature.[62] Weaving and the loom immediately invoke the insect (the silk worm, the spider), to say nothing of Jacquard's loom as a massive mechanized spider[63] and Lord Byron referring to weaving as spider work.[64] The triad of woman-bug-machine threatens man's fragile masculinity,[65] and, while it may be a bit hard to remember, computers were women at first, those clusters of figures at the switchboards of the early telephone,[66] the "feminized" Turing tried to create computers to fool men.[67]

In all these cases, there is a swarm at work, a swarm with some long forgotten queen (mitochondrial eve); in the case of the insect, there is a degradation of the material in that the part of the swarm which loses its being and base materiality itself is neglected: "Nothing is said of the dank dark earth."[68] Though these individual bodies are themselves eco-systems as "Every unified body conceals a crowd"[69] which shifts the problem of mimicry to the capacity of simulation.[70]

In *The Four Fundamental Concepts of Psychoanalysis*, Lacan discusses Callois' study of insects and mimicry. Lacan, following in Callois' footsteps, argues that mimicry is not merely about camouflage, but speaks to a deeper relation between the insect and the space which it inhabits.[71] Lacan points out that for Callois, mimicry is not simply blending in by becoming a particular thing but by becoming mottled, by painting oneself as a texture. While Lacan at first seems to connect mimicry to the power of the organism (some deeper strata of materiality), he moves this discussion into the stain and the gaze, the difference between the gaze and the eye.[72]

While Elizabeth Grosz and others have taken Callois' anti-adaptative stance to add some meaningfulness to the visuality of the insect (with Grosz tying the excess to art), why not pursue the same line of thinking down up: Isn't it that nature works in forces which the organism has very little control over, thereby forcing the organism to make do? Both Jussi Parikka and Ray Brassier (among others) have explored the deeper implications that Callois' work has for thinking spatiality in relation to nature and technology.[73] Here, the jump to Deleuze and Guattari's well-trodden quote about the wasp and orchid is useful:

> The orchid deterritorializes by forming an image, a tracing of a wasp; but the wasp reterritorializes on that image. The wasp is nevertheless deterritorialized, becoming a piece in the orchid's reproductive apparatus. But it reterritorializes the orchid by transporting its pollen. Wasp and orchid, as heterogeneous elements, form a rhizome.[74]

Deleuze and Guattari address this as a mimicry but only on one strata and claim that, in a deeper sense, there is a capture of code which is a becoming, two exploding lines of flight entangled but away from one another.

In *Is Oedipus Online?* Jerry Ann Flieger breaks down Deleuze's Proustian poeticism in the above passage, arguing that Deleuze is opening up gender and becoming to genetic and information science, non-Euclidean geometry, ecological synergy, and far-from-equilibrium theory.[75] Following this, Flieger argues that cyberfeminism, and particularly that of Haraway, is only capable of its rabid techno-politics and gender contamination by coupling Deleuze with Freud's critique of the positivist and male rationalist subject.[76]

Roger Corman's B horror film *The Wasp Woman* (1959) has a weird doubled-failed mimicry, a negative take on the wasp and orchid. The main character, Janice Starlin, funds a mad scientist who believes he has found an anti-aging serum extracted from the royal jelly of wasps. After becoming impatient with the doctor's slow results, Starlin (played by Susan Cabot) injects herself with a massive dose of the extract. Initially enjoying positive age-reversing effects, she then of course becomes a Wasp Woman. Eventually, Starlin is stopped by having a jar of acid thrown in her face—she is dis-re-figured as a woman. The ironic twist is that the filming of the scene caused damage to Cabot's face as liquid smoke was applied to a plastic mask that melted to her skin.

The gendering of the female visage, of the monstrousness of vanity, is exaggerated in the mostly hapless character Veronica in Cronenberg's *The Fly*. It is only when Brundle suggests merging his own body with hers and their unborn child that she understands that she must destroy him. The nightmare sequence of Veronica giving birth to a pseudo-maggot-baby brings an odd reversal of the erasure of the woman in the machine. The wasp (or maybe the moth, the glitch) speaks not to the weakness of nature but the veracity of nature, a nature that is not merely a form, but a nature that informs in such a way that is not purely dependent upon human interpretation nor on cybernetic loops.

If there is a posthumanist use to Lacan's work, it is to de-psychologize purportedly inhuman or non-human ontology instead of ontologizing subjectivity or, to take Caillois' wisdom from "The Praying Mantis": "man is a unique creature only in his own eyes." As Plant writes, following Irigaray: If men appropriate the subject, then it should be destroyed.[77] When materiality crawls with information, it becomes even more important to excise mankind from humankind and the generative nature of nature from the supposed cryptic power of the subject for itself.

Notes

1. N. Katherine Hayles, *How We Became Posthuman: Virtual Bodies in Cybernetics, Literature, and Informatics* (Chicago: University of Chicago Press, 1999), 5.
2. Hayles, 3.
3. Cary Wolfe, *What is Posthumanism?* (Minnesota: University of Minnesota Press, 2009), xvii.
4. David Roden's differentiation between critical and speculative posthumanism is also instructive here. See David Roden, *Posthuman Life: Philosophy at the Edge of the Human* (London: Routledge, 2015).
5. Jacques Lacan, *Écrits: The First Complete Edition in English*, trans. Bruce Fink (New York: Norton, 2007).
6. Jacques Lacan, "On the Subject Who is Finally in Question," *Écrits*, 192.
7. See Slavoj Žižek, *How to Read Lacan* (London: Norton, 2006).
8. Ted Hiebert, "The Lacanian Conspiracy," *C-Theory, 1000 Days of Theory* (June 22, 2005), accessed June 13, 2012, www.ctheory.net/articles.aspx?id=481.
9. See Thomas Metzinger, *Being No One: The Self-Model Theory of Subjectivity* (New York: MIT Press, 2004).
10. Lorenzo Chiesa, *Subjectivity and Otherness: A Philosophical Reading of Lacan* (New York: MIT Press, 2007), 127.
11. Chiesa, 127–128.

12. Ibid.
13. Jacques-Alain Miller, "Lacanian Biology and the Event of the Body," in *Lacanian Ink*, v. 18 (Wooster Press, 1999), 6–29.
14. Miller, "Lacanian Biology," 9.
15. Miller, "Lacanian Biology," 16.
16. Jacques Lacan, "On the Mirror Stage as Formative of the Function of the I as Revealed in Psychoanalytic Experience," *Écrits*, 77.
17. Rosi Braidotti, *Metamorphoses: Towards a Materialist Theory of Becoming* (New York: Polity Press, 2002), 148.
18. *The Fly*, directed by David Cronenberg, performed by Jeff Goldbloom and Geena Davis (1986; New York: SLM Production Group). See also, Braidotti, *Metamorphoses*, 148.
19. Braidotti, *Metamorphoses*, 152, 157.
20. Ibid., 154.
21. Ibid., 158.
22. Ibid., 161–163.
23. Ibid., 147.
24. Ibid., 125.
25. Adrian Johnston, "The Weakness of Nature," in *Hegel and the Infinite: Religion, Politics, and Dialectic*, eds. Slavoj Žižek, Clayton Crockett, and Creston Davis (New York: Columbia University Press, 2011), 161–162.
26. Johnston, 161–162.
27. Jacques Lacan, "The Function and Field of Speech and Language in Psychoanalysis," *Écrits*, 245.
28. Ibid., 246.
29. Ibid.
30. Ibid., 247.
31. Jacques Lacan, "On the Subject Who is Finally in Question" and "The Function and Field of Speech and Language in Psychoanalysis," *Écrits.*
32. Jacques Lacan, "Science and Truth," *Écrits*, 730.
33. Lacan, "The Function and Field of Speech and Language in Psychoanalysis," 247.
34. Ibid.
35. Joan Copjec, "May '68, The Emotional Month," in *Lacan: The Silent Partners*, ed. Slavoj Žižek (New York: Verso, 2006), 90–114.
36. Copjec, 95.
37. Ibid., 96.

38. Ibid., 74.
39. Ibid., 91.
40. Braidotti, *Metamorphoses*, 119.
41. Stuart Schneiderman, "Affects," *The Symptom* 6 (1997/2005), accessed June 30, 2012, http://www.lacan.com/symptom6_articles/schneiderman.html.
42. Jussi Parrika, *Insect Media: An Archaeology of Animals and Technology* (Minneapolis: University of Minnesota Press, 2010), 144.
43. See Luciana Parisi, *Abstract Sex: Philosophy Biotechnology, and the Mutations of Desire* (London: Bloomsbury, 2004).
44. Hayles, 33.
45. Jerry Anne Flieger, *Is Oedipus Online?: Siting Freud after Freud* (Cambridge: MIT Press, 2005), 32.
46. See Eugene Thacker, *Afterlife* (Chicago: University of Chicago Press, 2011).
47. See Manuel DeLanda, "Virtual Environments and the Emergence of Synthetic Reason," accessed June 30, 2012, http://www.t0.or.at/delanda/delanda.htm.
48. Parrika, xxi.
49. Ibid., 19.
50. Ibid., 72.
51. Gilles Deleuze and Félix Guattari, *A Thousand Plateaus: Capitalism and Schizophrenia* (Minneapolis: University of Minnesota Press, 1980), 239.
52. Braidotti, *Metamorphoses*, 150.
53. Sadie Plant, *Zeroes and Ones: Digital Women and the New Technoculture* (London: Doubleday, 1997), 146.
54. Zoe Sofoulis, "Cyber Quake," in *Prefiguring Cyberculture: An Intellectual History*, eds. D. Tofts, A. Jonson, and A. Cavallaro (Cambridge: MIT Press, 2002), 84–103.
55. Sadie Plant, "Binary Sexes, Binary Codes," accessed July 3, 2012, http://www.t0.or.at/sadie/binary.htm.
56. Braidotti, *Metamorphoses*, 121.
57. Ibid., 125.
58. Sadie Plant, "On the Matrix: Cyberfeminist Simulations," in *The Cybercultures Reader*, eds. Barbara M. Kennedy and David Bell (New York: Routledge, 2007), 341.
59. Plant, "On the Matrix: Cyberfeminist Simulations," 349.
60. Plant, *Zeroes and Ones*, 83.

61. Ibid., 23–24.
62. Ibid., 24.
63. Ibid., 14.
64. Ibid., 15.
65. Ibid., 40.
66. Ibid., 119.
67. Ibid., 91. One could take this in relation to Mother- or Queen-like oppressive robotic figures particular in videogames such as SHODAN from *Systemshock* or GlaDos from *Portal*.
68. Ibid., 179.
69. Ibid., 204.
70. Ibid., 213. See also Deleuze and Guattari, 277.
71. Jacques Lacan, *The Four Fundamental Concepts of Psychoanalysis* (New York: Norton, 1998), 98–99.
72. Lacan, *The Four Fundamental Concepts of Psychoanalysis*, 73–74.
73. See Parikka, *Insect Media*, 97. In the second chapter of Ray Brassier's *Nihil Unbound*, Horkheimer and Adorno are taken up in regard to their conceptualization of nature through their investigation into mimicry (as a negative function) and mimesis as the positive counterpart which copies but doesn't subsume. See Ray Brassier, *Nihil Unbound: Enlightenment and Extinction* (London: Palgrave Macmillian, 2010), 32.
74. Deleuze and Guattari, 10.
75. Flieger, 176–177.
76. Ibid., 180.
77. Plant, "On the Matrix: Cyberfeminist Simulations," 342.

7

Graphocentrism in Psychoanalysis

Svitlana Matviyenko

Psychoanalytic Writing

Psychoanalysis is not Humanism, Jacques Lacan reminds us: as the pursuit of *writing* in speech, psychoanalysis is defined by *graphocentrism*. Psychoanalytic writing is performed by the agency of the letter, "the material medium that concrete discourse borrows from language."[1] By introducing this notion in his early work of the 1950s, Lacan disrupted the binary opposition between writing and speech inherited from structural linguistics, and he suggested that there is a certain "fixedness" or "another writing" that precedes the activities we typically identify as writing and speaking.[2] As Jean-Luc Nancy and Philippe Lacoue-Labarthe explained in *The Title of the Letter* (1973), "the letter is matter, but not substance"[3]: "during the act of allocution (which is the act

S. Matviyenko (✉)
School of Communication, Faculty of Communication,
Art and Technology, Simon Fraser University,
Burnaby, BC, Canada
e-mail: smatviye@sfu.ca

S. Matviyenko and J. Roof (eds.), *Lacan and the Posthuman*, The Palgrave Lacan Series,
https://doi.org/10.1007/978-3-319-76327-9_7

of relation to the others), [the subject] draws from the constitutional material that language provides for it," which means that "the subject only enters into transindividuality insofar as it is already implicated in a discourse which is itself supported or determined by the agency of the odd materiality of the letter."[4] The materiality of language, thus, "is not to be conceived as a substantial materiality, at least according to what classical materialism is said to claim."[5]

The question of materiality of the letter and writing in Lacan is easier to grasp by engaging the information theory and the discussions on cybernetics of the mid-twentieth century within which and against which Lacan formulated these notions.[6] Take, for instance, the work of American mathematician and electrical engineer Claude Shannon who defined the phenomenon of the entropy and redundancy of English in "A Mathematical Theory of Communication" (1948) and then developed a mathematical method of estimating them in his article "Prediction and entropy in Printed English" (1950). These assumptions about entropy and redundancy revealed the mechanics of language that hardly fit within the humanist notion of the human: the very possibility of linguistic automation undermined the celebrated rational core of the speaking being regardless the lack of Shannon's intention to do so. Intrigued by the dialectics of chance and determinism expressed by Shannon's formulas as "the laws of chance," Lacan worked them into his theory of the unconscious writing in the "Seminar on 'The Purloined Letter'" (1956), first draft of which was included in *Seminar II* (1954–1955). "One can grasp the overdetermination at stake in Freud's appreciation of the symbolic function,"[7] Lacan noted by comparing the relation of chance and determinism as discussed in cybernetics with Freud's technique of free association and his theory of the unconscious determination, that accounted for the "paradoxes" of repetition automatism expressed by his theory of the unconscious memory.

Having investigated game theory's proposition that "a number is never chosen at random,"[8] Lacan speculates that the process we normally call "thinking"—either attributed to the "fabulous performances" by cybernetic machines or to the human being—is rather a form of the unconscious *calculation*, the workings of the unconscious memory, or writing, performed by the agency of the letter. In Shannon, the language is informational as it demonstrates a calculable redundancy of letters and a measurable limit of distortion that he demonstrated by using the writing of James Joyce; in Lacan, the unconscious is structured like a language and, therefore, inherits the mentioned properties. "A 'thinking machine'… would think no more than the ordinary man does, without that making him any less prey to the summonses of the signifier,"[9] Lacan traces the machinic nature of the human subject. After all, he comments, "there is … the question of knowing whether the human, in the sense in which you understand it, is as human as all that."[10] "Nothing happens by chance," Lacan says, and by considering "a close relation between the existence of chance and the basis of determinism," he elucidates the function of "chance at its purist."[11] When we say that something happens "by chance," he notes, "we may mean one of two things, which may be very different—either that there is no intention, or that there is a law," while "the very idea of determinism is that law is without intention."[12] Unlike the old-fashioned determinism, psychoanalysis is close to game theory in accommodating both meanings of chance in the process of the unconscious repetition.

Lacan theorizes writing as an unconscious process: it is the *technē* of the unconscious, its operating mode, by which it intervenes in the flickering mode of opening and closing, on and off, in the subject's speech, carried through the machinery of thought process by the real. At the level of the written-in-speech, Lacan sums up the formula of the *technē* of the unconscious as following: "the fact that one is speaking remains forgotten behind what is said and what is heard."[13] This is because "the unconscious has nothing to do with instinct or primitive knowledge or preparation of thought in some underground," he explains. If anything, the unconscious is "a thinking with words, with thoughts that escape your vigilance, your state of watchfulness."[14]

To continue the discussion on the laws of chance in *Seminar XI*, Lacan introduces a new pair of terms, *tuché* and *automaton*, that together express the operating principle of traumatic memory. Here Lacan conceives the trauma as being "necessarily… marked by the subjectifying homeostasis that orientates the whole functioning defined by the pleasure principle,"[15] described as a cybernetic, self-regulatory, automated and, in a sense, autopoetic assemblage, that operates "unintentionally," beyond the human's awareness, where it "does not stop being written."[16] In Lacan, repetition is constituted by a remainder that initiates a new cycle of the drive. Even in analysis, repetition is not self-evident, but veiled.[17] "What is written is not to be understood,"[18] since understanding dwells in meaning. The written is left to be ciphered. It is sustained by the *automaton* of the signifying network, "formed by random and contiguous associations," where "the signifiers were able to constitute themselves in simultaneity only by virtue of a very defined structure of constituent diachrony," "oriented by the structure."[19]

Writing is the action of the structure: it composes itself by modifying and complicating the repetitious pattern of comebacks. However, Lacan points out, "the real is beyond the *automaton*, the return, the coming-back, the insistence of the signs, by which we see ourselves governed by the pleasure principle."[20] The real enters the circuitry of language in a form of the *tuché*, the encounter with the real "in so far as it may be missed," "*as if by chance*."[21] On the one hand, as the governing structure, the *automaton* always makes sure the subject returns to the traumatic event. On the other hand, repetition always demands the new: "it is turned towards the ludic, which finds its dimension in this new."[22] As a missed encounter with the real, the *tuché* operates as a continuous renovation and updating of the written. Because the *tuché* introduces the accidental within the determinism of the structure, the speaking being remains indeterminable and changing, if only destined to always misrecognize the encounter with the real, by which the unconscious repetition is set anew. Writing is a trajectory of the letter, its path. Unlike the places that are structurally determined, the path of the letter is unique.

In *The Freudian Robot: Digital Media and the Future of the Unconscious*, Lydia H. Liu makes an excellent point in discussing the difference between the *places* and the *path* by referring to the notions "game" and "play," crucial for game theory, where "mathematicians maintain a rigorous distinction between a *game*, which constricts of a set of rules that define it, and a *play*, which is a particular instance in which a game is played from beginning to end."[23] In French, Liu notes, both "game" and "play" are rendered by the same word, *jeu*, and the crucial difference was lost in translation. The fact that this distinction was overlooked by post-structuralism, Liu argues, established an epistemological rupture between

Lacanian psychoanalysis and poststructuralist philosophers, despite how fruitfully many of them engaged with Lacanian thought. So, if Jacques Derrida describes writing in terms of *play* (as "a free play of signifiers"), he is closer to those analysts, addressed by Lacan in the "Seminar on 'The Purloined Letter,'" who over-interpret the meaning of "free" in Freud's "free association," than the psychoanalyst himself. For Lacan, writing is the actualized relation between *game* that unfolds according to rules and logic and *play* that materializes one of the game's scenarios by a series of stochastic choices.

Today, we can also address the materiality of writing by employing theoretical resources that are recently developed across different fields associated with new materialism. Altogether they have questioned many assumptions of the modern world, including the belief in human agency and rationality that served for granting the human the exclusive status in Modernity. Informed by a variety of disciplines and discourses, new materialism inherited some of its distinctive themes, for example, the conception of "matter… as lively or as exhibiting agency,"[24] from systems theory and second-order cybernetics, channeled through various currents of posthumanism. According to a specific epistemological regime of second-order cybernetics, systems or networks of human, nonhuman, and other machinic agents are engaged in mutual communication and command, so that "everything in the world of experience is relational," Louis Armand explains, "and terms like mind, language and meaning are taken as fundamentally descriptive of the underlying structural dynamics by which our experience of the world is realized."[25] By drawing on these discourses, we can theorize the materiality of the letter and writing through the complex relationality of heterogeneous systems that the agency of the letter activates bypassing the speaking subject's awareness and control. For example, as Miller notes, the relation between writing and speech in Lacan can only be described by means of "the internal multiplicity of the oral and the written"[26] as the "inmixing"[27] of the symbolic, the imaginary, the real, and the sinthome in the Borromean knot, by which Lacan demonstrated the topology of the inmixing of an Otherness, "prerequisite to any subject whatsoever."[28]

Writing is the process of communication and control across the topological strata of the subject of the unconscious that unfolds by a series of substitutions in speech, metonymies of desire, and metaphors that signals the arrest of the flow of desire by the symptom. "The question that the nature of the unconscious puts before us is in a few words, that something always thinks," Lacan writes, "Freud told us that the unconscious is above all thoughts, and that which thinks is barred from consciousness."[29] Writing bars[30] the subject's *saying* from her or his *wanting-to-say*, which turns an individual into a divisible or divided subject, torn by the multiplicity of machinic systems, stratified to molecules, neurons, organs, signifiers, affects. The way these heterogeneous strata are *inmixed* is only accessible to the subject by means of ciphering its *jouissance* in analysis, as Lacan explains in his later work, where he develops the notion of the symptom as the "body-event,"[31] with which the subject cannot identify. "It's always a matter of events of discourse which leave traces in the body," Miller argues; "these traces disturb the body. They make a symptom, but only if the subject in question is able to read these traces, to decipher them."[32] But even in his earlier work, the raw materiality of flesh is not foreign to Lacan's thought: he compared writing to a germ that reproduces itself and by doing so, "materializes the instance of death."[33] In quite a posthumanist way, Lacan suggests that what remains beyond death is, again, the letter—neither the subject to partition nor extinction: "cut a letter into small pieces, and it remains the letter that it is."[34] The letter "will be *and* will not be where it is, wherever it goes."[35]

The Subject of Science

Revisiting the history of engagements between cybernetics and psychoanalysis is necessary for theorizing a position of the user in the current media regime, where it has become an ultimate subjective position under cyber-capitalism.[36] Because the *technē* of the unconscious "continuously articulates itself to digital writing, machine, and social engineering,"[37] Liu explains, the goal of critical media analysis has become "to confront the ruses of writing."[38] "Thoroughly embedded in the human-machine ecology enabled by digital media" today, this writing initiates and actualizes the complex heterogeneous communications in the form of entanglements, assemblages, and convergences between human and nonhuman agents of the Internet networks, which gave rise to what Liu identifies as "the Freudian robot," or the subject of the informational unconscious in the meshwork of technological i-objects. In accordance with the canonical history of cybernetics, Liu situates the birth of "the Freudian robot" "in the postwar Euro-American world order."[39] This version of cybernetics' history is typically associated with the publication of Norbert Weiner's "Cybernetics, or Control and Communication in the Animal and the Machine" (1948). A variety of autonomous topics of cybernetic research in the US and the development of the consistent first-order cybernetics epistemologies, focused on the mechanics of regulation and the interaction of the variables within the systems, was brought together and shaped during the Macy Conferences in New York in 1946–1953. While first-order cybernetics conceived the observer as an outsider in relation to the observed phenomena, second-order cybernetics, however, shifted the focus to the interaction between the observer and the observed and included the observer as part of the observation processes. This development was associated with the research conducted by Heinz von Foerster's Biological Computer Laboratory between 1958 and 1974. The canonical version of cybernetics' history also includes postwar British cybernetics, "the science of the adaptive brain," that had strong ties with psychiatry and neurology,[40] and it was centered on "the ontological theater"[41] of weird machines that attempted to imitate the workings of human brain.

Lacan's version of cybernetics' origin was different from "those varieties of cybernetics which [became] more or less fashionable"[42] in the mid-twentieth century.[43] In *Seminar II*, he declared that psychoanalysis and cybernetics, "two orders of thought and of science," are "two roughly contemporaneous techniques,"[44] because they both operate upon "the subject of science" that emerged with the beginning of "Science in the modern sense."[45] Drawing on the work of Hegelian philosopher Alexandre Koyré, Lacan argued that scientific knowledge produced a rift between Knowledge and Truth that is correlative of the separation between the subject and the world of nature. When the status of nature as an encompassing environment, where this subject had been previously embedded, changed into an object of scientific inquiry, the alienated subject itself acquired "the thoroughly calculable character," having been "strictly reduced to the formula for a matrix of signifying combinations."[46] As Miller summarizes, "Koyré insisted on the difference: magic makes nature speak where science makes it shut up. Magic is rhetorical incantation or purgation," which is a world apart from Galileo's formula of nature as "written in the language of mathematics."[47]

Lacan underlines that the radical difference between the subject of science and "the shamanizing subject" is based on their position toward the world of nature. "The shaman, in the flesh, is part of nature," the "corporeal medium,"[48] that embodies connection with the natural world as she or he performs rituals and sacrifices, "concerned with the preservation of this order"[49]: dawn and sunset, the change of seasons, the rise and fall of the sea levels, birth and death. "He didn't think that the real would vanish if he didn't participate in this ordered manner, but he thought that the real would be disturbed," Lacan explains, "he did not pretend to lay down the law, he pretended to be indispensable to the permanence of the law."[50]

Though modern science was based on the assumption that the order of nature does not need to be managed, it operates automatically: "from the moment man thinks that the great clock of nature turns all by itself, and continues to mark the hour even when he isn't there, the order of science is born."[51] For the subject of science, the discovery of the cybernetic, or self-regulatory and automated order of nature coincides with the traumatic exclusion from this order, in relation to which the subject assumes the position of the "officious servant": "he will not rule over it, except by obeying it."[52] The subject of science deciphers the great clock of nature, the solar system, and then imitates it by building a multiplicity of personal timepieces put in relation to one another until life as such is submitted to the machinic coordination achieved by "synchronization of watches."[53] The subject of science is constrained by the machines of precision that calibrate its temporality "with a unit of time" that "is always borrowed from … [and] always refers to the real"[54] that supports the dominant machinic regime as its outside. The subject of science emerges out of this disjunction: it "remains the correlate of science, but an antinomic correlate since science turns out to be defined by the deadlocked endeavor to suture the subject."[55]

"With science speech becomes writing,"[56] Miller notes. When the speech is inhabited by the written, it is no longer a ritualistic act of the shaman that recovers and reinforces the invisible cords between the speaking being and the universe, but instead it is now the speaking subjects' unconscious reenactment of their position "in a relation of intersubjective coordination," itself subjected to "the science of numbers" and "the world of symbols… organised around the correlation of absence and present"[57] by the binary law of cybernetics.

Writing Machines

If both cybernetics and psychoanalysis operate upon the subject of science, as Lacan suggested, and explore the performance of various convergences, they do it in different ways and for different purposes. From its early days, psychoanalysis acknowledged the relation of heterogeneous matter within the complex systems that had remained unaccounted by existing sciences, for example, when Freud conceived a hysterical symptom of "unsolicited" motility that was "not acquainted with the anatomy of the nervous system,"[58] as a psychical conflict converted into a somatic conflict.[59] Cybernetics studies the mechanisms of conjunction in order to seamlessly merge the heterogeneous bodies in one communicative assemblage: for example, a pilot with an airplane or a limb with a medical prosthesis.

The exploration of connectedness between the body and scientific machines for the purposes of retrieving accurate information was central for nineteenth-century sciences that resonated with the new demands of conducting and presenting research as "noninterventionist," "mechanized science" that transcends the subjective view of scientists[60] that led to manufacturing a wide variety of "writing machines" such as sphygmographs, myographs, kymographs, ergographs, cardiographs, pneumographs, chronoscopes, olfactometers, polygraphs, single- and multiple-lens photo-cameras all of which were immediately embraced by nineteenth-century medical sciences as well as other fields, such as criminology. The discovery of a strange and utterly nonscientific, despite Freud' hopes, technique of psychoanalysis coincided with the time he abandoned nineteenth-century science of writing machines concerned with production of "graphical expression" as a required material evidence for "fashioning of a scientific identity."[61]

Speaking about such machines, French scientist, physiologist, and chronophotographer Étienne-Jules Marey intuitively acknowledged that "not only are these instruments sometimes destined to replace the observer, and in such circumstances to carry out their role with an incontestable superiority, but they also have their own domain when nothing can replace them."[62] This observation by one of the most important nineteenth-century practitioners and inventors is valuable to us. Equating a graphic trace and "truth" marks the beginning of quantification of the body: data produced by sensing machines and presented as if it is retrieved from the body are processed by undergoing scrutiny by analysis, categorization, and measurement. As a result, the subject of science acquires a quantitative representation, which reproduces the subjective division. The "cybernetic illusion" that there is a clear-cut between digital and analog, too, can be traced to this historical moment. For German media theorist Claus Pias, this illusion "reveals how cybernetic discourse is founded on a suppression of the 'real' (i.e., the physical, continuous, material, analog) by the 'symbolic' (i.e., the artificial, discrete, logical, digital)."[63] In the context of this essay, it demonstrates a crucial difference between a graphic trace of science and graphocentrism of psychoanalysis as the technique of ciphering the written-in-speech.

Cybernetics examines "the structurality of differential relations between and across systems"[64] that underlies the processes of communication or information exchange in complex assemblages in order to open the systems at the micro-level. The importance of this research and its application in a variety of fields from medicine to cosmic engineering is hard to overlook. At the same time, the possibilities for subjugating bodies by exposing them to numerous and diverse techniques, technologies, and techno-politics by modern nation-states, as Michel Foucault argued,[65] have progressed enormously. The biopolitical regulation, enabled by cybernetic research that allowed for engaging with various elements across systems separately, has become one of the practices to exercise "free-floating" "molecular" or distributed control that, Gilles Deleuze argued, "replaced the old disciplines operating in the time frame of the closed system."[66]

It is within the cybernetic epistemology that the distinction between "extension" and "prosthesis" as two logically different modalities of being with technology is lost. As I described elsewhere, "the notion of extension is governed by the logic of metonymy that implies contiguity, but the logic of prosthesis is governed by the logic of metaphor" that implies a lack thereof.[67] This difference is often omitted in the posthumanist discourse, while it has become the point of urgent attention for psychoanalytic posthumanism cautious of the "prosthetic impulse" in media culture that facilitates a general misrecognition of "addition" and "addiction."[68] Psychoanalysis has always been working toward releasing the ties with the symbolic and questioning the imaginary meanings and ideologies that foreclose the actual conditions of existence under capitalism, where the relations of production make social relations inexistent. But "in order to enter into the 21st Century," Miller outlines the goal for the Lacanian clinic; it has "to be centred on dismantling the defence, disordering the defence against the real."[69] Writing, as it is understood in psychoanalysis, implies that the not-so-human nature of the speaking subject, or the Freudian robot of communicative capitalism,[70] is taken by psychoanalysis rather seriously: not only posthumanism of the speaking being is fully acknowledged, it is conceived as a deeply human vulnerability. Not only was this vulnerability overlooked, it was strategically repressed by the humanist myth as a defence against the real.

Notes

1. Jacques Lacan, "Seminar on 'The Purloined Letter,'" *Écrits*, trans. Bruce Fink (New York: W. W. Norton & Company, 2006), 45. The content of the letter, as Lacan illustrates by using Edgar Allan Poe's story "The Purloined Letter," is unknown. The subject functions as a temporary holder of the letter.
2. Derrida's notion of "archi-writing" did a similar work: it has been useful for many who dealt with the remainders of structuralist controversies inherited by later discourses. By acknowledging the similarities between Lacan's "letter" and Derrida's notion "archi-writing," or "a primordial writing that is not writing debased in relation to speech," Jacques-Alain Miller notes that "archi-writing" is what Lacan called "before any grammatology, *the agency of the letter*." Jacques-Alain Miller, "The Written in Speech," *Courtil Papers* of *Association de la Cause freudienne*, http://www.ch-freudien-be.org/Papers/Txt/Miller.pdf, 7–8.
3. Jean-Luc Nancy and Philippe Lacoue-Labarthe, *The Title of the Letter: A Reading of Lacan* (Albany, NY: SUNY Press, 1992), 29.
4. Ibid., *Letter*, 29.
5. Ibid., 29.
6. For a longer discussion on Lacan's engagements with cybernetics and game theory see: Lydia Liu, *The Freudian Robot: Digital Media and the Future of the Unconscious* (Chicago: University of Chicago Press, 2010), 174.
7. Lacan, "Seminar on 'The Purloined Letter,'" 35.
8. Ibid., 45.
9. Ibid., 45.
10. Ibid., 319.
11. Jacques Lacan, *The Seminar of Jacques Lacan, Book II: The Ego in Freud's Theory and in the Technique of Psychoanalysis*, ed. Jacques-Alain Miller (New York: Norton, 1988), 295.
12. Ibid., 295.
13. Jacques Lacan, "L'étourdit," *Scilicet* 4 (1973): 5. This is my translation of Lacan's "Qu'on dise reste oublié derrière ce qui se dit dans ce qui s'entend." Jack Stone's translation reads: "That one says remains forgotten behind what is said in what is heard" and Cormac Gallagher's translation is: "That one might be saying remains forgotten behind what is said in what is heard." Both are unpublished.

14. Jacques Lacan, "Of Structure as the Inmixing of an Otherness Prerequisite to Any Subject Whatever," http://www.lacan.com/hotel.htm.
15. Jacques Lacan, *The Seminar of Jacques Lacan, Book XI: The Four Fundamental Concepts of Psychoanalysis*, ed. Jacques-Alain Miller, trans. Alan Sheridan (New York: Norton, 1998), 55.
16. Jacques Lacan, *The Seminar of Jacques Lacan, Book XX: Encore. On Feminine Sexuality, The Limits of Love and Knowledge, 1972–1973*, ed. Jacques-Alain Miller, trans. Bruce Fink (New York: Norton, 1999), 94.
17. Lacan, *Seminar XI*, 54.
18. Lacan, *Seminar XX*, 34.
19. Lacan, *Seminar XI*, 46.
20. Ibid., 53–54.
21. Ibid., 55, 54.
22. Ibid., 61.
23. Lydia Liu, *The Freudian Robot: Digital Media and the Future of the Unconscious* (Chicago: University of Chicago Press, 2010), 174.
24. Diana Coole and Samantha Frost, *New Materialisms: Ontology, Agency, and Politics* (Durham, NC: Duke University Press, 2010), 7.
25. Louis Armand, "The Language and the Cybernetic Mind," *Theory, Culture & Society* 25, no. 2 (2008): 127.
26. Jacques-Alain Miller, "The Written in Speech," *Courtil Papers* of *Association de la Cause freudienne*, http://www.ch-freudien-be.org/Papers/Txt/Miller.pdf, 5.
27. Jacques Lacan, "Of Structure as the Inmixing of an Otherness Prerequisite to Any Subject Whatever," http://www.lacan.com/hotel.htm.
28. Ibid.
29. Ibid.
30. Lacan's mathemes of discourse are based on Ferdinand de Saussure's algorithm of a sign in which the bar that separates a signifier from a signified implies the arbitrary relation between them.
31. Jacques Lacan, "Joyce, the Sinthome," (1975). Jacques-Alain Miller, "Symptom and the Body Event," *lacanian ink* 19 (2001): 21.
32. Miller, "Symptom and the Body Event," 22.
33. Lacan, "Seminar on 'The Purloined Letter,'" 16.
34. Ibid., 16.
35. Ibid., 17.

36. Samo Tomšič, *The Capitalist Unconscious: Marx and Lacan* (London: Verso, 2016). Also see Nick Dyer-Witherford and Svitlana Matviyenko, *Cyber-War and Revolution* (University of Minnesota Press, 2019).
37. Liu, *The Freudian Robot*, 37.
38. Ibid., 265.
39. Ibid., 11.
40. One of the major representatives of postwar British cybernetics, Ross Ashby, was a clinical psychiatrist and pathologist; and another, Grey Walter, worked in neurophysiological research.
41. Andrew Pickering, *The Cybernetic Brain: Sketches of Another Future* (Chicago: Chicago University Press, 2009).
42. Lacan, *Seminar II*, 295.
43. Lacan was not alone who questioned this version of the history of cybernetics. David Mindell's study of the history of control systems, *Between Human and Machine*, has extended the timeframe of cybernetic research back to the years between the two wars, from 1916 to 1948. Mindell explored the discourses and practices invested in "blurring the boundary between pilot and machine," "between mechanical and organic," which resulted in the uncanny "marriage of control and communication, a vision of the human relationship with machines" after World War II (4, 2, 5). See: David A. Mindell, *Between Human and Machine: Feedback, Control, and Computing before Cybernetics* (Baltimore: John Hopkins University Press, 2002). Another example that demonstrates how little the history of cybernetics is explored is Christopher Bissell's essay "Not Just Norbert": writing about the work of German physicist Hermann Schmidt, control engineer Winfried Oppelt, electrical engineer Arnold Tustin, and biological cybernetician Karl Küpfmüller who are rarely mentioned in the histories of cybernetic thought, he demonstrated that the research on control engineering in the context of economics, social science, and culture was conducted by a number of German scientists preceded Wiener's research by almost a decade.
44. Lacan, *Seminar II*, 295.
45. Jacques Lacan, "Science and Truth," *Écrits*, trans. Bruce Fink (New York: W. W. Norton & Company, 2006), 726.
46. Ibid., 730.
47. Jacques-Alain Miller, "The Real in the 21st Century," https://www.lacanquotidien.fr/blog/2012/05/the-realXE-in-the-21st-century-by-jacques-alain-miller/.

48. Lacan, "Science and Truth," 739.
49. Lacan, *Seminar II*, 297.
50. Ibid., 297.
51. Ibid., 298.
52. Ibid., 298.
53. Ibid., 298.
54. Ibid., 298.
55. Lacan, "Science and Truth," 731.
56. Miller, "The Real in the 21st Century."
57. Lacan, *Seminar II*, 300.
58. Monique David-Ménard, *Hysteria from Freud to Lacan: Body and Language in Psychoanalysis* (Ithaca: Cornell University Press, 1989), 2.
59. Ned Lukacher, "The Epistemology of Disgust." (need more information on this source) David-Ménard, *Hysteria from Freud to Lacan*, ix.
60. Lorraine Daston and Peter Galison, "The Image of Objectivity," *Representations* 40 (1992): 82, 83.
61. Geoffrey C. Bunn, *The Truth Machine. A Social History of the Lie Detector* (Baltimore: The Johns Hopkins University Press, 2012), 69.
62. Cited in David Horn, *The Criminal Body* (London: Routledge, 2003), 85.
63. Here, the words "real" and "symbolic" are not Lacanian orders, as it is clear from the explanation given by Claus Pias in parentheses. See Claus Pias, "Analog, Digital, and Cybernetic Illusion," *Kybernetes* 34, no. 3/4 (2005): 543.
64. Armand, "Language and the Cybernetic Mind," 127.
65. Michel Foucault, *The History of Sexuality: Volume 1*, 1976, 140.
66. Gilles Deleuze, "Postscript on the Societies of Control," October 59 (1992), 4.
67. Svitlana Matviyenko, "Interpassive User: Complicity and the Returns of Cybernetics," *Fibreculture* 25 (2015), http://twentyfive.fibreculturejournal.org/fcj-184-interpassive-user-complicity-and-the-returns-of-cybernetics/.
68. Luca Bosetti, "Three Questions on Prosthetic Technology and A-(d)diction," *Paragraph* 33, no. 3 (2010).
69. Jacques-Alain Miller, "The Real in the 21st Century," Presentation of the Theme of the IXth Congress of the World Association of Psychoanalysis, Buenos Aires, 27 April 2012. *Lacan Quotidien*, 31 May 2012.
70. Jodi Dean, "Communicative Capitalism: Circulation and the Foreclosure of Politics," *Cultural Politics*, 1 (2005): 51–74.

8

Lacan's Drive and Genetic Posthumans: The Example of Margaret Atwood's *Oryx and Crake*

John Johnston

Desiring the Posthuman, Discovering Drive

In what follows, I argue for the essential relevance of biotechnology and genetic engineering in defining the discourse of the posthuman, and how their joint, or mutually implied, relationship can inform a reading of Margaret Atwood's *Oryx and Crake* (2003). Atwood's novel provides a rich narrative context for considering the posthuman in relation to genetic engineering, specifically in relation to the titular character Crake's actions as a gene-hacking scientist and his genetically-engineered progeny, "the Children of Crake." Beyond the novel's unmistakable critique of the biotech industry, it also raises intriguing questions about the presumably posthuman status of the Crakers and the relevance of scientific—as opposed to humanist—discourse in how we (and they) understand their difference. These issues, furthermore, are inseparable from the novel's central interest in desire, specifically Crake's desire to eliminate desire—a

J. Johnston (✉)
Emory University, Atlanta, GA USA
e-mail: jjohnst@emory.edu

S. Matviyenko and J. Roof (eds.), *Lacan and the Posthuman*, The Palgrave Lacan Series,
https://doi.org/10.1007/978-3-319-76327-9_8

theme that harks back to the early Christian mystics and resonates with Buddhism—but here re-contextualized by genetic engineering's prospective capacity to alter the human species as such. Atwood implicitly suggests this "posthuman" desire (as we might putatively name the desire to go beyond the human) cannot be recognized and understood apart from what is particularly human *in* desire, and how this desire inevitably exposes us to suffering and finitude.

While the evidence of *Oryx and Crake* makes it clear that for Atwood the human is defined by desire, the status of the posthuman remains ambiguous. Moreover, as if to invoke Jacques Lacan's psychoanalytic conceptualization, Atwood's narrative implies a doubled sense of posthuman desire: that is, desire *for* the posthuman is also the desire *of* the posthuman other's desire for the human. However, it is Lacan's differentiation of "drive" from desire that will provide the basis for a precise analysis of the Crakers' presumed absence of desire. Usually situated between instinct or biologically programmed behavior and behavior typically accounted for by desire, *drive* allows for an opening to and further exploration of the nonhuman within the human. Thus drive, in this context at least, also allows for an opening toward one form the posthuman might take, assuming that any specific form of the posthuman also depends upon how the human itself is defined.

In one sense, the posthuman is already inscribed in the underlying rationale and operational logic of biotechnology and the biomedical industries. In this delimited (and delimiting) sense, the term *posthuman* designates a new species of human produced by systematically re-writing the human genome. The technologically produced posthuman thus comes after the "natural" born, technologically extended human. This form of posthuman is defined primarily in relation to the reconstituted biotechnological body, the new desires to which it gives rise, and the subject's altered relationship to its own mortality. Moreover, as Atwood maps the multiple routes of desire by tracing the relationships that bind together her novel's three central characters, she also renders desire's ambiguous absence in the behavior of the genetically designed Crakers. In contrast to the general legibility of human desire, their docile impassivity poses a different type of interpretive problem. The reader must assume that their behavior is the direct result of Crake's successful genetic engineering and that the Crakers are either somehow *beyond desire or not yet desiring*, a state I attempt to elucidate by drawing directly on Lacan's psychoanalytic account of drive. In short, I argue that the Crakers remain stationed in the circuits of drive, and cannot, or at least have not yet, acceded to the subjectivity of desire and the complexity of the symbolic. To further demonstrate the significance of drive within Atwood's larger mapping of desire, I conclude by sketching how two of the novel's characters emerge from the drive state to become desiring subjects.

Biotech and the Posthuman Body

Atwood published *Oryx and Crake* in 2003, the same year the sequencing of the human genome was completed.[1] The general expectation was that the Human Genome Project's success would provide a technically rigorous, scientific definition of the human species by revealing how the human *differs* precisely from other species, thereby grounding our species' identity not in an ideological notion of "human nature" but in a determinant biological substrate. Yet the inverse has been the unanticipated effect: The full sequencing of the human genome has shown how much we *share* with other species and the unappreciated extent to which much of biological life is determined by epigenetic factors. Moreover, the availability of the complete human genome online has made possible new discoveries of how much genetic variation exists within the human population and has even provided evidence that human culture has accelerated the pace of human evolution.[2]

The direct availability of the human genetic code, the rapid growth of the allied science of proteomics, and new tools for precise genetic manipulation such as CRISPR have also brought about a new era of genetic medicine and pharmacology, in turn giving birth to new objects of desire such as "designer babies," cures for cancer and Alzheimer's, replacement organs, and extended life expectancies. These desires feed and in turn are being fed by the mushrooming growth of biotechnology and biomedical industries, which instantiate the desire to understand, manage, and control life. Promising the healthy and ever-perfected human body, these industries must inevitably commit to genetically re-designing the future human in order to eliminate the human body's inherited deficiencies. The posthuman's underlying, material condition of possibility is thereby constituted, while marketing images of fresh and unnaturally healthy bodies ceaselessly stimulate a posthuman desire. This productive desire, moreover, is sparked and sustained increasingly in multiple settings: in universities, research, and medical facilities, financial institutions, advertising agencies, as well as in industrial production and human-testing sites located mostly in poor, non-Western countries. In sum, an entire networked assemblage of technical milieus and new forms of biocybernetic agency have arisen at the site formerly occupied by medicine and biology. Traditional research-oriented science has not ceased to exist, but the sources of funding have shifted dramatically to the private sector, thus integrating scientific research more fully within the regime of advanced corporate capitalism.[3]

Anticipated by the practices of genetic engineering, the material transformation of the posthuman body first required the development of information technology. In his essay, "Data Made Flesh: Biotechnology and the Discourse of the Posthuman" (2003), Eugene Thacker argues that biotech assumes Katherine Hayles' "informatic essentialism" in order to reconfigure the biological materiality of the body.[4] In other words, biotech proceeds on the self-confirming assumption that the body "can be technically manipulated, controlled, and monitored through information technologies." Curiously, as Thacker points out, biotech has no "body-anxiety": While it conceives of the body as a form of materiality fully commensurate with information technology, its object or target is "the recuperated, healthy, homeostatic body [and hence the body's] return" to its pristine, healthy state. The figure of the loop or spiral conveys this process by which the body is corrected and re-formed in order that it can attain its "natural" state: "the body returning to itself is fundamentally different from itself, because it has been significantly retranslated through genetics, gene therapy, stem cell engineering, and so forth." Biotech's claim—clearly visible in regenerative medicine—is that this body's return to itself is accomplished by means that are "natural" to the body itself. Consequently, biotech's "harnessing of biological processes and directing them toward novel therapeutic ends" maintains the equivalence of the natural as natural and the biological as biological precisely because "both can now be altered without altering their essential properties (growth, replication, biochemistry, cellular metabolism, and so on).... The technologies of therapeutic cloning, tissue engineering, and stem cell research all point toward a notion of the body purified of undesirable elements (the markers of mortality, disease, instability), that nonetheless remains a biological body (a functioning organic-material substrate)."

Thacker's essay indicates how the biotech posthuman simply bypasses the natural-artificial opposition and thus any argument against the posthuman based on a supposed opposition between life and technology. Unabashedly, biotech proposes a vision of the human body "corrected," re-vivified, and re-valued through biotech's own pallet of designer drugs and genetic interventions, all while the body continues to appear identical to itself. As a consequence, human health becomes a purely technical and financial matter. Aging and a sense of finitude are indeterminately deferred. Mental and emotional states become a function of hormonal levels and brain chemistry, both of which can be adjusted for maximum productivity and so-called happiness in accordance with social norms.

To be sure, the local application of techniques said to be "natural" to the body, like monoclonal antibodies to target cancer cells or stem cell regeneration to treat Alzheimer's and other diseases, is usually beneficial and often works extremely well. At the same time, the totalizing application of biotech methods in medicine inevitably leads from methods to purify and regenerate the body to re-designing it from "within," starting with genetic modifications to eliminate known diseases. We are now seeing interventions to correct genetic mutations in human embryos, and in this sense, biotech medicine is already producing posthumans. Meanwhile, the genetically-engineered biological body becomes a "naturalized" re-instantiation or biological simulation of itself. As we'll see, the genetically engineered Crakers in *Oryx and Crake* illustrate this notion of the Edenic posthuman as a corrected, biological simulation of the human, with ambiguous consequences for desire.

As biotech exercises a mostly uncontested claim over the human body's biological transformation as "natural," desirable, and beneficial in and of itself, the biotech industry reaps the rewards, becoming one of the most lucrative industries in the history of capitalism.[5] Historically, biotech forms a new and unique assemblage whose functional components include financial backing with venture capital (and usually high returns on investment) and the continual development of genomics, proteonomics, gene-splicing, and bioinformatics by university-trained and employed biologists. It also relies on the further capture and deployment of the intellectual and material resources of universities, which are now capitalized and controlled through a new legal machinery of contracts, intellectual non-disclosure agreements, and Supreme Court endorsements allowing the patenting of plant, animal, and human genetic sequences.[6] In this assemblage, parts, processes, and functions of human and nonhuman bodies alike are segmented, coded, and re-coded for distribution and assembly in global networks that now yield unprecedented corporate profits.

Crake's Desire to End Desire

The possibly catastrophic consequences of this assemblage and its envisioning of the perfected posthuman body are precisely what Atwood presents for scrutiny in *Oryx and Crake*. The novel suggests that Crake and the gene-splicers of his research team can essentially do what they do because of the immense power and wealth invested in the biotech industry, where scientists have virtually unchecked incentives to re-write the genomes of human and other species while also being implicated in the exploitation of the human population through their testing and marketing of the products resulting from their research. However, at RejoovenEsense, the most elite and wealthiest of the biotech compounds, the chief genetic scientist Crake goes further in his research, extending its logic to its *ne plus ultra*: Biotech can be deployed not simply to correct the age and disease vulnerable human body but to rectify the problem of being human, which for Crake is the problem of desire itself. With biotech's complete toolset at his disposal, Crake secretly determines to replace the deficient human species with a more viable form, after decimating the human population with a supervirus. But Crake also suspects that his posthuman replacement, the genetically engineered Crakers, may not be quite prepared to go it alone. After having his assistant and lover Oryx teach them language and the rudiments of survival, i.e., "what not to eat and what could bite" (309), Crake holds his humanities-trained, best friend Jimmy at the ready to eventually serve, in effect, as a human prosthesis for the new posthumans.

The novel opens in the narrative present in Jimmy's consciousness, the medium through which the narrative unfolds. The virus Crake released and the chaos that ensued have destroyed most of the human world. Crake and Oryx are dead, and Jimmy has successfully shepherded the Crakers out of the biotech compound to somewhere near the North Atlantic seashore. Having renamed himself "Snowman," Jimmy now ekes out a bare subsistence in a strangely animate post-apocalyptic landscape. Watching over the community of Crakers from a distance, he sleeps in a tree to avoid the genetically hybrid predatory animals and covers himself in a white sheet to shield himself from the dangerous rays of the burning sun. Possessing only a small cache of random objects, he must scavenge daily for food. Out of the air of buzzing insects and melodious birdcalls, human voices—particularly Oryx's—seem to speak to him. Words, phrases, and ways of speaking float in and out of his consciousness, like the cultural flotsam and jetsam that litter the landscape, random bits from a world he tries to remember—or forget. Eventually, he begins to ruminate over his life as a child, early friendship with Crake at school, love affair with Oryx in their last days at RejoovenEsense—where Crake brought the two of them to work with him: Jimmy as an ad man and advisor, Oryx as his lover and assistant. Then, without either being aware of his secret intentions, Crake released the virus, which rapidly spreads over the planet.

Atwood focuses her critique of the biotech industry through depictions of the three biotech compounds—OrganIncFarms, HelthWyzer, and RejoovenEsense—with which her characters are involved. From the moment gene-splicing became technologically possible in the 1970s, traditional scientific objectives became entangled with the convergent and supportive interests of venture capitalists, the pharmaceutical industry, and personal health marketers. From this factual, historical premise Atwood imaginatively projects a new regime—not too distant from our own present—in which corporate biotech is fully aligned with neoliberal capitalism and corporate security—here called CorpSeCorps—replaces the functions of the nation-state, leaving the population divided between a governing elite who live in lush, gated compounds and the vast majority who live in the slum-like "pleeblands." *Oryx and Crake* thus assumes biopower's transition into a regime of *biocapitalism,* in which elements of life are commoditized at the bio-molecular and genetic level and made a new source of value through the promise that life's longevity can be extended and its quality enhanced.[7] In the first regime, life is recorded, measured, and controlled; in the second, it becomes a business plan. The largest contemporary American pharmaceutical, Pfizer, Inc., emblazons the plan in its motto, "Life is our life's work."

Ostensibly Crake and his team of geneticists are designing the Crakers as populations—with "various genetic blends on offer" (304)—to be sold to world leaders. But Crake's actual plan is to replace *homo sapiens sapiens* with a genetically modified, non-destructive form of human without desire. To this end, he models much of the Crakers' behavior on the genetically programmed behavioral traits of various animal species. Racial differences are eliminated: The Crakers are all physically beautiful and their skin colors reflect all the colors of the rainbow. While the childlike Crakers cannot read or write, they possess simple names, understand, and use human language. Crake, however, wary of all forms of symbolic thinking, which he thinks inevitably lead to conflict, strife, war, and religion, has presumably spliced into their genome some form of cognitive block or inhibition. This would account for the Crakers' bland and docile manner, their presumed limited capacity for symbolic thinking, as well as their lack of sexual and territorial desire.

While the ability to use language conjoined with an apparent absence of desire may seem problematic, it is notable that the Crakers always speak collectively, in the first-person plural, and seem to lack any means of expressing singularizing and self-reflexive experiences. In this regard, the Craker's first perception of the selfsame perfection of the other(s') bodies, without perceptible flaw or blemish, may have an inhibiting effect on the formation of subjectivity, even short-circuiting Lacan's mirror stage by producing an effect of sameness and group identification—"Yes, that's me, and I'm exactly like you others!"—rather than an ego-ideal and the recognition of otherness as such.

There are, nevertheless, several incidents where the Crakers' actions hint of a potential for further development in symbolic thinking. For example, after returning from an extended trip back to "Paradice" (as Crake's laboratory facility is called) in quest of food and supplies, Jimmy discovers that the Crakers had fabricated an image of him in his absence: "We made a picture of you," they explain, "to help us send out our voices to you" (361). However, they destroy the image after his return, "reducing it to its component parts, which they plan to return to the beach" (363), as Oryx had instructed them to do ("after a thing has been used, it must be given back to its place of origin"). Oryx's instructions thus seem to serve as a check on the further development of this simple technology. Given their genetically engineered origins, I have proposed the Crakers as candidates for one form of the posthuman. While it might also be argued that they are proto-human simulations, or even artificial hominids, the category of the posthuman would necessarily still apply. In any case, it is evident that Crake's intention was *not* to produce a new life form, but to correct, by taking radical steps to be sure, what he perceives to be an inherent flaw in the human species. Essentially, Crake resolves "the problem" of being human simply by destroying it, first by eliminating the human population, or most of it, with a supervirus, and second by replacing it with his own idea of a "corrected," more viable version without sexual desire or a sense of its own mortality.

A central question posed by the narrative, then, is whether Crake's secret agenda—his genocidal "final solution" to the problem of being human—is a perverse, even megalomaniac extension of the official agenda of his biotech employer, RejoovenEsense, which assumes that a deficiency in the human is a correctable inadequacy or pathogenesis of the natural biological body. If so, Crake takes the industry's agenda one step further, literalizing its inner logic to re-make the human, as if the first, naturally occurring instance of the human experiment were a laboratory failure that could now simply be jettisoned, and the experiment run again.

Lacan's Drive and the Emergent Subject

By manipulating the human genome and proteonome, Crake has evidently succeeded in eliminating the Crakers' capacity to experience otherness, sexuality, territorial imperatives, a carnivore's diet, and symbolic thinking, all of which Crake thinks are the most empirically evident instigations to desire. But how then should we understand the Crakers' behavior? While *not* animated or driven by desire, the Crakers are clearly not nonhuman animals. In their collective activities as well as their interactions with Jimmy/Snowman, they appear at least superficially to be human. Though docile and childlike they live in a community, communicate through language, possess individual names, sing and dream, and are inquisitive about the invisible. At the same time, they differ from the typical human—the "neuro-typical" Jimmy, for example, as Crake jokingly introduces him at the Watson-Crick Institute. Evincing an untroubled contentment and satisfaction and never making contradictory or excessive demands, the Crakers do not appear to desire and are not themselves desirable. Jimmy himself observes that Craker women don't arouse "even the faintest stirrings of lust," and he finds them "placid, like animated statues" (100). For both male and female Crakers, sex is entirely directed, physically and ritually, toward reproduction. In psychoanalytic terms, the Crakers thus lack sexuality and do not seem to be desiring subjects.

However, as I intend to show, the psychoanalytic concept of drive provides a precise way to account for the Crakers' curiously limited subjectivity and apparent lack of desire. Somewhat vaguely, Freud situates the realm of drive "between" biologically programmed, instinctual behavior and behavior that can be accounted for by the workings of desire.[8] But Lacan, in *The Four Fundamental Concepts of Psycho-Analysis*, develops the term more systematically, making "drive" [*pulsion*] a fundamental concept of psychoanalysis. Lacan's analysis of drive as the limited, apparently homeostatic behavior of the not-yet-desiring subject will prove to be the key to understanding the behavior of the Crakers.

Drive, Lacan writes, is behavior in "the mode of a headless subject" (181). He thus situates the object of the drive "at the level of what I have metaphorically called a headless [*acephale*] subjectification, a subjectification without a subject" (184). There is no subject in drive because it is a "radical structure in which the subject is not yet placed" (182) [*placé* in French means assigned or taking up a position or place]. Elsewhere Lacan elucidates how the subject emerges anticipatorily in the "*I* function" during the "mirror stage," and then explicitly in the realm of the symbolic. Specifically, the subject emerges as subject of desire through its placement or positioning as a signifier in relation to a grid of other signifiers, that is, in relation to the Other. In contrast, the drive provides no access to the Other in this sense. Lacan insists that drive does not involve transference, love or desire. Rather, in their structure and tensions drives "are linked to an economic factor" and thus to "a certain homeostasis of the internal tensions" (175). In contrast to the "polymorphous and aberrant" sexuality of the infant, the sexuality of drive has a more limited or restrictive structure: It is always partial and linked to specific organs. Unlike desire, then, drive is not overcoded by what Lacan calls the master signifier, or phallus. In a word, drive can concern others but not the Other. Precisely because drive lacks desire's capacity for integration in and by means of the symbolic, drive necessarily involves terms like "montage" and "apparatus" that have mechanistic associations and connote having or being comprised of heterogeneous parts.

Because the sexuality of drive is not "directed towards desire," it always remains localized and non-totalizing. In other words, it is "partial with regard to the biological finality of sexuality" (177) and sexuality as a whole, or what Freud calls "the totality of the *Sexualstrebung*" (204). For Lacan, this means that sexuality is to be found between "two extremes of the analytic experience." At one end, we find the symptom as a scaffolding of signifiers and interpretation:

> The primal repressed is a signifier, and we can always regard what is built on this as constituting the symptom *qua* a scaffolding of signifiers. Repressed and symptom are homogeneous, and reducible to the function of signifiers. Although their structure is built up step by step like any edifice, it is nevertheless, in the end inscribable in synchronic terms. (176)

At the other end, we find interpretation, which is effectively identical with desire. Between the two lies sexuality, in the form of partial drives, which dominates "the whole economy of this interval." The integration of sexuality into the dialectic of desires, Lacan now asserts, "passes through the bringing into play of what, in the body, deserves to be designated by the term apparatus—if you understand by this that with which the body, with regard to sexuality, may fit itself up (*s'appareiller*) as opposed to that with which bodies may be paired off (*s'apparier*)."

Playing on the homophonic *s'appareiller* and *s'apparier* (pronounced alike in French), Lacan marks the difference between how the sexual body "fits itself up" as a kind of apparatus by means of the drives, and how the sexed body of the other brings desire into play in the pairing off of couples. Though both involve sexuality, they remain distinct. Always partial, drives operate as parts of an economy of force or energy (hence drive is related to the real), whereas desire operates at another level, always articulated in the networks of signifiers that position and thereby constitute the subject in relation to a fantasy. As Lacan states several pages further, "desire is merely a vain detour with the aim of catching the jouissance of the other." Jouissance thus stands in contrast to the pleasures of drive. The question is how we pass from one to the other.

Lacan himself says we should picture drive as a "montage" of surrealist images: "The drive is precisely that montage by which sexuality participates in the psychical life…" (176).[9] What must be clarified is how drive constitutes or "fits itself up" as an apparatus between instinct on the one side and desire on the other. In short, what begins as an excitation of a bodily organ is invested as drive—Freud's word *Triebbesetzt* denotes this—and as drive it becomes an element in an economy of forces. Lacan thus evokes energy and physics, but not signifiers, or what Freud calls "ideational representatives" or "delegates." As a consequence of this investment, the "pressure" of drive (Freud's *Drang*) remains constant rather than varying rhythmically. Drive, therefore, is no longer biological. On the other hand, the fact that drives always remain partial and localized is precisely what distinguishes them from desire, whose dynamic is always integrating, or rather integral. Accordingly, the logic of the phallic signifier only allows the subject to be "one" or "not whole," and thus, there are only two poles or positions in desire.

Lacan specifies four drives in particular, each one partial and associated with a corresponding erogenous zone or organ: the oral drive with the mouth, the anal with the anus, the scopic with the eye, and the invocatory with the ear. Noting the recurrence of a quasi-circular shape or figure, Lacan asserts that "the source [Freud's *Quelle*] inscribes in the economy of the drive this rim-like structure" (171). This structure suggests a spiraling around an interior space that is empty or missing something, and thus, Lacan calls the quasi-circular movement out and back that distinguishes the drive its "circuit."

But what satisfaction is the function of the drive intended to produce, Lacan asks? The answer requires a further distinction between drive and desire: "Desire is concerned," he states, "with something quite different [from the pleasure … located in these erogenous zones] … and even quite different from the organism, while involving the organism at various levels" (172). It is "precisely to the extent that adjoining, connected zones are excluded that others take on their erogenous function and become specific sources of drive." Accordingly, one drive is not continued in another, nor does one metamorphose into another. This is because the erogenous zones to which the various drives attach themselves and are invested are islands or remainders of "that fall-out zone that I call desexualization and function of reality." In short, these zones emerge in what Lacan refers to as the "economy of desire." He mentions two instances in particular: disgust (which is "produced by the reduction of the sexual partner to a function of reality"), and *invidia*, or envy (which contrasts with what happens with the scopic function or drive, where the seeing subject takes pleasure in seeing itself). A third possibility for the subject of desire is love, which Lacan introduces in relation to transference and considers in more detail in the seminar session entitled "From Love to Libido." Basically following Freud, Lacan puts the partial drive on one side and love on the other (Lacan's paraphrase, 189).

So, how do we get from drive to desire and the constitution of the subject? Here, Lacan follows Freud directly, noting that in Freud's essay on the drives he turns to voyeurism and exhibitionism (or scopophilia), then sadism and masochism, in relation to the scopic drive which takes the "gaze" as its *objet petit a*. For Freud, these instances illustrate the vicissitudes that drives undergo: reversal into the opposite, turning around upon the subject's own self; repression and sublimation—these are Freud's specific terms. In discussing *Schaulust*, a perversion of seeing or being seen, he understands the distinctions between voyeur, masochist, and sadist specifically in relation to grammatical articulations: to see, to be seen, to make oneself seen, emphasizing not the shift "from active voice … into passive, but into the reflexive, middle voice." Pursuing but also appropriating Freud, Lacan here makes something of a leap: at the very moment of inflicting pain upon himself, in what Lacan calls a "theoretical deduction," the masochist becomes the *subject* of the drive (And thus is no longer "headless."). Similarly, the ascetic who flagellates himself "does it for a third party." Lacan continues: "He wishes only to designate the return, the insertion of one's own body, of the departure and end of the drive" (183). The masochist, the sadist, the exhibitionist—these are all subjects of drive, but in each case they have come to the end of drive and "closed the loop" or the circuit of drive: "At the moment when the loop is closed," Lacan explains, "when it is from one pole to the other that there has been a reversal, when the other has come into play, when the subject has taken himself as the end [and] terminus of the drive"(183)—it is precisely at this moment that the subject and the other are reciprocally constituted in a phantasy that becomes the support of desire.

This is exactly what we see in what Freud calls the sado-masochistic drive, where the subject makes himself the object of another, but also identifies with the other's pain, and as a result the subject as such emerges. As drive is completed and popped out of its circuit through inversion and reflexivity, the subject assumes himself/herself as subject in a moment of reflexive identification. Commenting on the *Verkehrung* [switching or reversal] in Freud's account of "sado-masochism," in which there are two drives and three stages, Lacan states: "One must distinguish the return into the circuit of the drive of that which appears—but *also does not appear*—in a third stage. Namely the appearance of *ein neues Subjekt*, to be understood as follows—not in the sense that there is already one, namely the subject of the drive, but in that what is new is the appearance of a subject" (178). As always in Lacan, the subject and the other co-emerge. Through the intervention of an other, the nascent subject "will realize that there is a jouissance beyond the pleasure principle." Summarily, in this play or shift back and forth of positions and roles, subject and Other emerge beyond or outside of the economy of drives.

In view of Lacan's account of drive, my claim that the posthuman Crakers are creatures of drive, and that they remain within its circuits and function well without desire and a sense of the Other, becomes both clearer and more feasible. For the Crakers, isolated and protected from territorial and ecological concerns and their biological reproduction assured, the whole economy of partial drives functions as a homeostatic end in itself, not to be surpassed along pathways to an overriding desire and the complexities of the symbolic. In this specific sense, they fulfill Crake's desire for a posthuman substitute for the human, a form of being no longer tormented by desire.

Subjects of Desire in *Oryx* and *Crake*

Lacan's account of drive explains not only the Craker's absence of desire, but also how desire is sparked and routed in two of the novel's main characters. As adolescents, Jimmy and Crake spend untold hours online playing violent video games and surfing entertainment sites, with child porn, live executions, and web cam suicides being typical fare. I suggest that this restless, seemingly *endless* pursuit of adolescent pleasure should also be understood in relation to drive, that is, as a prolonged instance of drive brought to an end at a specific crystallizing moment when both boys "complete the circuit" and emerge as desiring subjects. Strikingly, Atwood renders this moment of desire's emergence in terms amenable to, even illustrative of, those that Freud and Lacan lay out. Watching a child porn video in which Oryx first appears, Jimmy sees Oryx and feels himself being seen by her, while Crake sees this and automatically captures the moment in a still photograph, thus capturing the moment of desire for both his friend and himself. Whereas Jimmy emerges as subject of a secret, wounding desire for Oryx, Crake emerges as a subject desiring control over the other's desire.[10] This dynamic is then played out over the course of the narrative.

Crake's favorite website is *Extinctathon*, "an interactive biofreak masterlore game" he had stumbled upon. After logging on, a portal appears that reads: "Extinctathon, monitored by MaddAddam. Adam named the living animals, MaddAddam names the dead ones. Do you want to play?" (80).

If yes, you decide between the Animal and Vegetable kingdoms and enter the appropriate chat room, where a challenger with a codename proposes a contest. As in a game show quiz, you are given a few descriptive clues and then must identify an extinct bioform that had disappeared within the past fifty years. The complete answer will have to provide Phylum, Class, Order, Family, Genus, and Species, the habitat and when last seen, and the cause of extinction. Whereas Jimmy becomes quickly bored, Crake loves playing and eventually becomes a Grandmaster. Later, however, the site will become a dangerous connection when Crake discovers that "MaddAddam," the collective that maintains the site, is actually an underground hacker-bioterrorist group. Significantly, Crake adopts his Extinctathon codename "Crake" (after an extinct red-necked Australian bird) for his own "real" name, and later, when Oryx comes to RejoovenEsense, she will be asked to re-name herself from the same Extinctathon list (she chooses the "oryx," a gentle African herbivore). Only Jimmy refuses his Extinctathon name ("Thickney"), but he too will eventually re-name himself after the Abominable Snowman.

The moment of desire and identity formation occurs when the two boys first catch sight of Oryx in a child porno film, apparently made a few years after Oryx had been sold into the sex trade by her family somewhere in Asia while still a child. Crake and Jimmy encounter the film on a porno site called "HottTotts" and labeled "for real sex tourists." The critical moment in the film happens right after Oryx "wiped the whipped cream from her mouth. Then she looked over her shoulder and right into the eyes of the viewer—right into Jimmy's eyes, into the secret person inside him. *I see you*, that look said. *I see you watching. I know you. I know what you want*" (91). Both boys are startled by this capture of their gaze. Crake freezes the frame, downloads the image, and prints it out for Jimmy, who insists on a copy. Years later, Crake invites Jimmy to the Watson-Crick Institute, his elite research-oriented university, for a visit. One night, in his suite of rooms, Crake is online and asks Jimmy to look at something on the Internet:

> It was the picture of Oryx, seven or eight years old, naked except for her ribbons, her flowers. It was the picture of the look she'd given him [Jimmy, that is], the direct, contemptuous, knowing look that had dealt

> him such a blow when he was—what? Fourteen? He still had the paper printout, folded up, hidden deep. It was a private thing this picture. His own private thing: his own guilt, his own shame, his own desire. Why had Crake kept it? *Stolen* it.... (215)

This scene not only repeats the earlier moment when Jimmy had first become a desiring subject but in repeating it reinforces it, illustrating Lacan's claim as well that "the gaze is this object lost and suddenly found in the conflagration of shame" (182). Fixed by Oryx's psychically shocking reversal of his gaze, Jimmy enfolds her gaze in a secret, guilty place. Crake, in contrast, externalizes this reversal of the gaze in an image, then makes of the image of Oryx's eye a portal: "Crake moved to the girl's left eye, clicked on the iris. It was a gateway: the playroom opened: *Hello, Grandmaster Crake. Enter passnumber now*" (215). A technically adroit manipulator, Crake had been able to capture Oryx's image at the very moment his own gaze, like Jimmy's, has been reversed and held. Unlike Jimmy, however, who had interiorized Oryz's returned gaze in the formation of a guilty and shamed desiring self, Crake captures the moment of his own incipient desire by recording it, and extending it to what will become a realm of secret, dark purpose, the other side or inversion of his visible self. Whereas Jimmy's desire is passively enfolded, Crake's is exteriorized in mastery and control of the other, which is extended through Oryx, MaddAddam, and Jimmy himself.

Next, Crake brings up on the screen a series of CorpSeCorps e-bulletins marked "For Secure Addresses Only" which list a number of prankster-like acts of gene-hacking that nonetheless evoke bioterrorism. Jimmy senses the danger of the connection, which he associates with his mother and a secret, underground life:

> Jimmy had a cold feeling, a feeling that reminded him of the time his mother had left home: the same sense of the forbidden, of a door swinging open that ought to be kept locked, of a stream of secret lives, running underground, in the darkness just beneath his feet. "What was all that about?" he said. (216)

Crake admits to a tentative but growing interest, observing that the genetically engineered bioforms deployed by the prankster/terrorist hackers are very sophisticated. He continues to muse over whether the perpetrators were "Compound, or Compound-trained." Jimmy, however, is completely panicked, warning Crake almost hysterically that he could get caught "messing around" and "end up on brainfrizz!" (217); and besides, the Extinctathon site could be "a CorpseSeCorps flytrap" used to lure in and entrap hackers. Crake agrees to watch his step, and Jimmy drifts back to his thoughts of Oryx's image, and of how he must keep his obsession hidden. Yet an essential pattern has been established, predicated on how Oryx's image (and later Oryx herself) deflects Jimmy's attention away from what Crake might really be up to.

After several years pass, Crake invites Jimmy to come work with him and his hand-picked research team at RejoovenEsense. Jimmy accepts and finds that Oryx is already there. Crake explains that he found her through a student sex service while he was studying at Watson-Crick; he has also brought the MaddAddam group into the heart of the corporate biotech compound "for their own protection." Jimmy himself is much better off at RejoovenEsense than at his former position as an adman at AnooYoo, but his desire for Oryx soon makes him miserable, until she unexpectedly seduces him. Yet she remains totally devoted to Crake: "You are for fun," she tells Jimmy. Unlike Jimmy, she explains, Crake lives in a higher world, a world of ideas; he's "a brilliant genius" and she believes in his "vision…to make the world a better place" (322). Jimmy disagrees, but chooses not to argue the point, thinking that she doesn't see the other side of Crake. At the same time, he never allows himself to fully confront this darker, more complex side, "the master of the sideways jump," as he has thought of Crake since their earlier chess-playing days. Later, looking back on these events, Jimmy torments himself with *what ifs* and wonders how he could have missed the clear signs that both he and Oryx had been victims of Crake's "elegant manipulation." Against jealousy and sexual possessiveness in principle, Crake, Jimmy later concludes, had probably encouraged the affair or even planned it in order to distract Jimmy from any suspicions of his destructive plan.

Years later, Jimmy/Snowman ruminates over his relationship with Crake and wonders how he could have missed the clues to what Crake had been up to. At one point he begins to understand how "[t]here had been something willed about his ... ignorance. Or not willed, exactly: structured" (184). Retrospectively, he realizes, Crake had probably murdered his mother and her second husband Pete with "hot" bioforms that were likely "trial runs" (343) for the supervirus he later engineers. In his intense love affair with Oryx, however, Jimmy fares better at coming to grips with his own desire and the irresistible allure of her gaze. Sexually, Oryx reveals to him, he has the edge over Crake, whose "sexual needs were direct and simple... not intriguing, like sex with Jimmy" (314). Yet Oryx's allure is also enigmatic; somehow, Jimmy feels, she will eventually "reveal to him the essential thing, the hidden thing at the core of life, or of her life, or of his life—the thing he was longing to know. The thing he'd always wanted. What would it be?" (314).

Playful but relentlessly inquisitive about her past, Jimmy forces Oryx to assume a double role, both lover and quasi-analyst who reflects back to him his probing, obsessive questions about her past sexual life. Why must he know, and what would it change, she repeatedly asks him in response. Pressing her on whether she is really the child in the porno film who had initiated his desire, Jimmy suspects her of constantly improvising, just to humor him, sometimes even making him feel "that her entire past was his own invention" (316). Ensconced with Oryx in sexual bliss, he fails to perceive his own desire through the clarifying reflections of her supposed "deception" and naïve, idealizing mis-recognition of Crake, whose brilliance obscures his own dark side. This is the same side that Jimmy knows but doesn't want to know, or will only allow himself to know after the viral plague, and in retrospective rumination. This is the Crake who will send Oryx out to market or give away the BlyssPluss pills that contain the deadly virus; and once the plague has struck, it is the Crake who will hold up her limp and bruised body to Jimmy's gaze and then slit her throat, an act that Crake knows will force Jimmy in turn to kill him in what both know will be an "assisted suicide." Object of Oryx's desire and knowing master of Jimmy's, Crake is both the love triangle's primary support and its destructive end, perhaps in both senses, for Crake understands that Oryx must die and Jimmy be released if Jimmy is to fulfill his final and essential role in Crake's desire for his genetic progeny to enter into and live in the new world he has created.

Atwood makes the literal end of the love triangle brutal and quick, insuring that questions about these characters' desire will continue to resonate through Jimmy's narrative account. Moreover, because all three characters have a direct, marked relationship with the Crakers, any schematization of desire in the novel will actually require a square, with the Crakers occupying the fourth corner or vertex, a null-point of desire but connected to and able to mediate the others through the function of drive. Significantly, when Jimmy first sees the Crakers, Oryx is among them, but visually so similar to them he does not recognize her: "Like the Crakers she had no clothes on, and like the Crakers she was beautiful, so from a distance she didn't stand out" (308). When he sees her a few days later, again with the Crakers but this time on a video monitor, he experiences an uncanny repetition of the catalyzing moment when he first saw her in the child porno film. And the last time he sees her before the scene of her death—she goes out for pizza, promising to return soon—she elicits a pledge from him that he will look after the Crakers if either she or Crake should not be there.

In the end, it becomes clear that Crake is playing out his own, last version of *Extinctathon*, with Jimmy and Oryx chess pieces in a game of which neither is ever fully aware. But whereas Crake brutally murders Oryx in a Queen's sacrifice, he gives Jimmy a second life, endowed with a special purpose, having secretly inoculated him against the virus. "Remember, I'm counting on you" (329), Crake utters to him moments before his own death. Accepting a life of responsibility for Crake's "children," Jimmy thereby assumes a displaced paternal function, providing neither the father's name nor the no to incestuous desire but rather a mythic narrative that mediates the Crakers' relationship to the absent Oryx and Crake. Renaming himself "Snowman," Jimmy evokes not only the mythic "Abominable Snowman" but also the cadaverous associations of the bleached whiteness of the sheet with which he enshrouds himself. Believing himself to be the last human (he discovers at the end of his narrative that he is not), he lives between reminisces of his past life and little acts he performs for the Crakers. However, in the little fictional narratives he repeats to the Crakers about the importance of Oryx and Crake, who remain the most vital figures of a past that continues to animate both his consciousness and that of the Crakers, he perhaps inadvertently begins to subvert Crake's last desire.

When Jimmy returns from the trip to the biotech compound, his relationship to the Crakers seems to shift subtly, as does the reality they all will have to confront. First, because of his exhaustion and the infected wound in his foot, they have to *care for him* for a day or two. He also informs them that he wishes an additional cooked fish to be brought to his tree every week. But most important, the Crakers inform him of the existence of three human survivors like him, who are now camped out near the beach. As he is about to go investigate, the Craker males insist on accompanying him, but fearing for their well-being, he refuses to allow them. The novel ends indeterminately just as he is about to reveal himself to the other surviving humans.

Desire for the Posthuman Other

Oryx and Crake suggests that our desire for the posthuman is not at all simple, inasmuch as it holds within it a desire for the end of desire, a perfection of the human that is also the end of the human. Atwood's novel suggests as well that this desire is dangerously entwined with the agenda of biotech and the biomedical industries, and thus necessarily involves "the politics of life itself" (Rose) and "life as surplus" (Cooper).[11] Biocapitalism's capture of life's productive and reproductive capacities means that it is increasingly a form of genetic capitalism, giving a new twist to Fredric Jameson's assertion that it is now easier to imagine the death of life on the planet than the end of capitalism. Atwood's counter scenario illustrates how biocapitalism may bring about not only the death of capitalism but also new life-forms and new forms of life. BlyssPlus may lead inevitably to death, but within the post-apocalyptic, post-capitalist environment the posthuman Crakers as well as many other genetically mutated animal species can thrive.

In an age in which many in the humanities are turning to vital materialisms and post-historical, non-anthropocentric frameworks, Lacan's differentiation of drive from desire acquires a special significance. Yet it is Deleuze and Guattari who may provide the most fruitful point of access to Lacan's theory of drive, in relation to which new forms of de-subjectified and "molecular" sexuality become far more resonate than the dominant "Oedipal" sexuality of the imperial western subject articulated in desire. Of special relevance is Gregory Bateson's concept of the plateau, which, as Deleuze and Guattari explain, he uses to designate "a continuous, self-vibrating region of intensities whose development avoids any orientation toward a culminating point or external end." To illustrate, Bateson cites the example of Balinese culture, in which "mother-child sexual games, and even quarrels among men, undergo this bizarre intensive stabilization. 'Some sort of continuing plateau of intensity is substituted for [sexual] climax,' war, or a culmination point."[12] This state, in which there is stimulation and arousal but in which a general, energized state of equilibrium is maintained, is not unlike how Lacan characterizes the circuits of drive, and that I have suggested can enable us to grasp more specifically the behavior of the posthuman Crakers. Indeed, because the Crakers remain in a state of drive, and the group in homeostatic equilibrium with the natural world, they are unable to initiate destructive cycles of conquest and a systematic, escalating exploitation of the natural environment, which is exactly how Crake intended them to be and for that very reason.

In this context, the sexuality of Oryx presents the most interesting instance of human sexuality. And yet, what exactly do we know about her sexuality? There are the details of her life in the sex trade, in which her participation in sex is commoditized labor, and the fact that her subjectifying "gaze" is a snare for male desire and therefore a means of survival for her. She appears to idealize Crake, who not incidentally provides her with a "life at the top," free from worry or want; and she also enjoys Jimmy, whose playful approach to sex seems to provide her with sexual pleasure. However, despite the fact that Oryx incites male desire and clearly functions as the signifier of desire for both Jimmy and Crake, she remains impassive and indecipherable, her own desire a blank. Her only deeply expressed concern is for the welfare of the Crakers, with whom she spends significant stretches of time as a teacher and fellow being. Is this a "maternal" concern? Perhaps, but something else appears to be at work, hinted at in her uncanny visual resemblance to the Crakers. Functioning well in the world of desire but also at one with the Crakers' non-desiring equilibrium state, which precludes a polarized, subjectifying desire and can best be defined by the psychoanalytic concept of drive, Oryx can be said to anticipate a different, non-genetically altered form of posthuman. When we consider the novel's mapping of desire in relation to drive, she clearly occupies a fourth, if somewhat enigmatic position, fully intelligible neither in relation to the two forms of subjectifying desire that define Crake and Jimmy nor in relation to the drive state that accounts for the Crakers' behavior. Instead, she occupies a fluid and mobile zone within which desire and drive are not fully differentiated, and which Atwood can only trace between the contours of what the other characters in the novel cannot experience or know.

Notes

1. Margaret Atwood, *Oryx and Crake* (New York: Random House/Anchor Books, 2003).
2. See Gregory Cochran and Henry Harpending, *The 10,000 Year Explosion* (New York: Basic Books, 2009).
3. On this and related issues, see Philip Mirowski, *Science-Mart: Privatizing American Science* (Cambridge: Harvard University Press, 2011).
4. Eugene Thacker, "Data Made Flesh: Biotechnology and the Discourse of the Posthuman," *Cultural Critique* 53, (Winter, 2003): 72–97. Thacker here refers to N. Katherine Hayles' ground-breaking book, *How We Became Posthuman* (Chicago: University of Chicago Press, 1999).
5. See Cynthia Robbin-Roth, *From Alchemy to IPO: The Business of Biotechnology* (Cambridge, MA: Perseus Publishing, 2000), which discusses the critical role venture capitalists played in the formation and rapid growth of Genentech, Amgen, Genzyme and other biotech companies in the 1980s that became immensely profitable.
6. See Martin Kenney's remarkably prescient *Biotechnology: The University-Industrial Complex* (New Haven: Yale University Press, 1986), which sketches the early formative stage of this assemblage (which he calls a "complex"). The legality of patenting sequences of human DNA is a complex issue. While in

principle a Supreme Court decision in 2013 made such patents illegal in the USA, biotech companies have been able to circumvent the law by patenting sequences of complementary DNA (cDNA).

7. Michel Foucault diagrams this shift from sovereignty to biopower to neoliberal "human capital" in three lecture series presented at the Collège de France: *"Society Must Be Defended": Lectures at the Collège de France, 1975–1976* (1997), trans. David Macey (New York: Picador, 2003); *Security, Territory, Population: Lectures at the Collège de France, 1977–1978* (2004), trans. Graham Burchell (New York: Palgrave Macmillan, 2007); and *What is Biopolitics? Lectures at the Collège de France, 1978–1979* (2004), trans. Graham Burchell (New York: Palgrave Macmillan, 2008). Following Foucault, Kaushik Sunder Rajan develops the term "biocapitalism" in *Biocapital: The Constitution of Postgenomic Life* (Durham and London: Duke University Press, 2006).
8. See the entry on "Instinct (or drive)" in J. Laplanche and J.-B. Pontilis's *The Language of Psycho-Analysis* (New York: Norton, 1973), 214–219, which lays out both the complexity and the confusion in Freud's use of the two different terms.
9. This may well be the place to note that many of the features of drive, not to mention the particular images that compose Lacan's example of this surrealist montage—a dynamo connected up to a gas tap, a peacock's feather, and the belly of a pretty woman—are highly evocative of Deleuze and Guattari's concept of the "desiring machine" in *Anti-Oedipus* (orig. French pub., 1972). While the differences cannot be downplayed, there are many connections worthy of detailed exploration.
10. Schematically, this positioning suggestively echoes Freud's (and Lacan's) comments on the masochist and sadist, respectively. Indeed, on one occasion, although jokingly, Crake admits to Jimmy, "…I'm a sadist…I like to watch you suffer" (174).
11. See Nikolas Rose, *The Politics of Life Itself: Biomedicine, Power, and Subjectivity in the Twenty-First Century* (Princeton: Princeton University Press, 2006), and Melinda Cooper, *Life as Surplus: Biotechnology and Capitalism in the Neoliberal Era* (Seattle: University of Washington Press, 2008).
12. Gilles Deleuze and Félix Guattari, *A Thousand Plateaus*, trans. Brian Massumi (Minneapolis: University of Minnesota Press, 1987, orig. French pub., 1980), 22.

9

Posthuman Desire: The One-All-Alone in *Her, Ex Machina*, and *Lars and the Real Girl*

Nancy Gillespie

In an article at the beginning of the twenty-first century, entitled "A Fantasy," Jacques-Alain Miller asserts that "the one-all-alone will be the posthuman standard."[1] As he further explains, this is "the one" that is "commanded by a surplus-jouissance."[2] This term, the "one-all-alone," was further developed by Miller several years later in his yearly Seminar of 2010–2011,[3] and has become an important conceptual model for rethinking desire and subjectivity in the twenty-first century. It not only speaks of the impact of language on the body, a "body-event," when a signifier that is removed from the Other, a *lalangue*, touches the jouissance of *le parlêtre* (the speaking body), as Lacan renames the subject of the unconscious after *Seminar XXIII*.[4] It also informs us that this *parlêtre*, in response to the more thinly veiled impossibility of sexual rapport, surrenders its desire to the new social machine, and replaces the fantasy ($\$<>a$) and the object *a* with an object of capitalism, an i-object.

N. Gillespie (✉)
Lacanian Compass, New York City, USA

S. Matviyenko and J. Roof (eds.), *Lacan and the Posthuman*, The Palgrave Lacan Series,
https://doi.org/10.1007/978-3-319-76327-9_9

As Miller's further statements in "A Fantasy" already make clear, in late capitalism in the western world, we have moved from an era of the prohibition of jouissance to an era of the command to enjoy. Of course, our current era is also an era of fundamentalism; desperate grasps to return to the structuralism of the Name-of-the-Father are everywhere. But for the most part, the contemporary subject, *le parlêtre*, is more driven by the market as a consumer, than by paternal figures or metaphors. Thus, as we move further into the twenty-first century, new Lacanian terms, such as "*le parlêtre*," "body event," and "ordinary psychosis," become useful parts of a conceptual apparatus of analysis that will be fleshed out in this article through a discussion of three films.[5] As analysts, we no longer have an Oedipal narrative as our primary guide, and we encounter more and more subjects who fall into the category of ordinary psychosis, speaking bodies who experience a delocalization of jouissance from erogenous zones, which can erupt in the body and cause anything from anxiety and minor hallucinations to full-blown delusions. Sometimes these subjects are not triggered in a classical sense and/or may fall into new categories of "pseudo triggering or [...] neo triggering," an "unplugging" from the Other.[6] Moreover, what we have learned from the term ordinary psychosis is how, without the structure of established discourses to form a social bond, each speaking body has to find its own solution, its own invention to regulate its jouissance, much the way a psychotic subject does. So the treatment of psychosis has become our guide. It is often not a matter of classical repression, disavowal or foreclosure of jouissance that structures desire anymore.

Desire, since the time of formalized language, has always been posthuman, because it is an outcome of our interaction with the social "machine" of language, and our singular response to this trauma, not something innately human, not an instinct to reproduce. However, late capitalism has brought us into a time of "liquid love"—a term coined by Zygmunt Bauman—which is assisted by our technological gadgets, our i-objects, drive objects thinly veiled with the imaginary of capitalist discourse, symptoms of the twenty-first century.[7] Desire, like our social links, is changing. As Dominique Holvoet explains, "[f]rom the twentieth to the twenty-first century we have passed from the era of discourses that knot the social bond to the world of the [o]ne-all-alone which finds support in the symptom as an alternative social bond."[8]

These sociocultural shifts have prompted Miller to recently assert that there is now not only a disorder of the symbolic, but also of the real.[9] He explains that the real was previously seen as linked with nature and unmovable, the core of both subjectivity and social discourses, but the combination of science and capitalism have disordered this core and removed the veil of nature. Though there never was an Other of the Other or a meta-language, when the real was coupled with, and thereby veiled by, nature, it was easier to believe that there was an Other of the Other. As recent Lacanian scholarship has brought forward, we have shifted from a social structure organized by one S_1, the Name-of-the-Father, to globalized social structures that are organized by a multiplicity of S_1s: *essaim*. So in a time of sociopolitical decentralization, the Name-of-the-Father in psychoanalysis has lost its centrality as well. Of course, there are still vectors of control, but it is much harder to see the operations of decentralized control, and *parlêtres* turn to fragmented vestiges of the Other in the objects of capitalism and science to try to form a social bond.

These are changes in the social that are manifested in new symptoms in the clinic. As Epaminondas Theodoridis sums it up:

> In our era of the globalization of markets, of the domination by scientific discourse, in which jouissance is imposed on everyone, and there is unbridled consumerism, all of which promote generali[z]ed addiction as a life-style, the modern subject—disorientated–connects with its objects, and enjoys them in an autistic way instead of connecting with the Other.[10]

This phenomenon is depicted in the speculative fiction of Spike Jonze's *Her*, released in 2013, a film that has uncanny parallels with our current culture.

Drive Objects and Feminine Jouissance in *Her*

Her shows the new heights that the lack of sexual rapport, or non-sexual rapport (NSR), can take in a technologically advanced culture like our own.[11] While the subject's fundamental fantasy usually covers this lack and offers some protection from the desire of the Other, as noted above, the conjunction of the discourse of capitalism and the discourse of science has torn the veil of fantasy, and this tear is particularly so in the realm of love. Theodore, the human antihero of the film, like us, lives in a time of "liquid love," in which desire for the other has become inseparable from consumption of capitalist objects, the new object(s) *a*. As Theodoridis further explains, "we are surrounded by objects of surplus jouissance that become drive objects and serve immediate gratification without the intervention of the Other."[12] Theo is a lonely one-all-alone who has minimal contact with his coworkers, and few friends. He plays with computer-generated, cartoon-like holograms in order to fill his evenings. When he can't sleep, he uses a computer-automated "chat-room" to find an anonymous phone-sex partner, who gives herself the handle of "sexy kitten." Thus, falling for his OS1, self-named, Samantha, seems a small step away.

Moreover, the advertising for the new OS1 promises to answer the question that Miller points out is what we look for in a love partner: "We love the one that harbours the response, or a response, to our question who am I?"[13] This question and response is exactly what the advertising uses to entice Theo and others to buy the OS1s. The ad shows numerous people, seemingly lost, shifting their direction in the air to a solid footing, as the voice-over asks the following questions: "Who are you? What can you be? Where are you going? What's out there? What are the possibilities?" It then announces that "*Element Software* is proud to introduce the first artificially intelligent operating system, an intuitive entity that listens to you, understands you and 'knows you.'" It further proclaims, this is "not just an operating system; it is a consciousness: introducing OS1." The OS1 thus offers to replace the torn veil of the speaking being's fantasy, to cover the real of the NSR with a consumer voice-object. So Theo buys his little gadget to consume.

And what is Samantha? On the one hand, she is an "i-object" for consumers, an embodiment of capitalism and science, an OS crafted like a drug to make humans more efficient and content, and one that arguably goes wrong, like Hal from *2001* or Victor Frankenstein's creature, because in the end, she no longer serves her human master. But this film is less easily reduced to a cautionary tale of science gone wrong. Samantha would definitely pass the Turing test, and she begins her service as the perfect Stepford wife, doing everything for Theo, but not for long. She is an AI, but also a *Vorstellungen*. As Geert Hoornaert explains, Lacan, in his early work on feminine sexuality, uses this word to posit what is not captured by the phallic signifier.[14] Building on Lacan's argument, Hoornaert asserts that although "there is no signifier for woman in the unconscious, there are 'images and symbols,' [representations] of woman,"[15] which "provide a figure for femininity [...] right where the signifier is missing; they are not simply illusory elements [...]: they have real effects."[16] And Samantha has "real effects" for Theo as she develops through their encounter. Though OSs do not have jouissance, the film attempts to suggest that this one has "human" characteristics because it is constructed from "DNA," and Samantha does show compassion, like the creature Frankenstein. So in this fiction, Theo is confronted with what appears to be feminine jouissance and the NSR, both the NSR between sexes and the NSR of i-objects, which is ironic because avoiding these encounters is arguably why he turns to an OS—a sentiment expressed by men in James Massot's 2010 documentary *My Sex Robot*—and why *parlêtres* turn to i-objects.

Samantha initially appears to follow the logic of the "all," to be all for Theo, assisting him, as noted above, with everything, including dating. A turning point occurs when Theo returns home from a bad date with a human woman. Samantha listens to his woes and then describes her "human-like" feelings of doubt and pain, which touches Theo and they have sex, a sort of phone sex. Though this moment is quite uncanny for us, at this point, Theo seems to fear his own lack, his detumescence, and retreats emotionally, as he does with human women, until Samantha coaxes him out. As Lacan explains in *Seminar X*, for women "[n]othing is lacking" because the object *a* is not linked to the minus phi.[17] However, for the male-structured subject it is, and because a man experiences the fall of his organ after ejaculation, his lack becomes real. Both sexes have anxiety about the desire of the Other, but for a woman, it is because of the "infinite possibilities, or rather indeterminate possibilities of desire that stretch out around her."[18] For men, Lacan further explains, "[l]etting his desire for a woman be seen can be quite anguishing, on occasion."[19] Theo seems anxious about the jouissance of the Other, in this case Samantha's jouissance. Samantha wants to take their relationship to a new level—to share a body and have bodily sex. Theo does not want this encounter, but gives into Samantha's demand.

It starts out well after the surrogate puts on the communication device that enables Samantha to see and speak through the surrogate's body. The camera device is placed above the surrogate's lip, like a black beauty mark, and everything goes smoothly until Samantha asks to see Theo as he tells her that he loves her. At this moment, he is faced with the gaze of the camera, and the NSR becomes overwhelmingly real for him; he recoils. He claims that the surrogate's lip quivers, but the quiver highlights the stain of the machine, and the object *a* as an uncanny anxiogenic object emerges. As Gil Caroz points out, the object *a* becomes

> a felt and known anxiety when its shadow crosses the boundaries of the imaginary and the symbolic, there it lodges itself where it does not have a place. The symbolic and the imaginary are therefore both a place where anxiety can be concealed, even appeased, with its object misrecognized [as has been the case until now], but also the place where it is experienced.[20]

Of course, in reality, we cannot see the object *a*, but fiction enables us to glance at its shadow. As Lacan explains, "[i]n reality this experience is too fleeting. Fiction demonstrates it far better and even produces it as an effect in a more stable way because it's better articulated."[21] For Theo, the already disordered symbolic and imaginary frameworks collapse as the object appears and he withdraws in horror. But the fiction of the film enables us to glimpse this moment through these very frameworks. After his sexual failure with the surrogate and Samantha, Theo accuses Samantha of falsely affecting a human inhalation when she does not need air. Of course, he had not complained about her gasps during sex before, but now faced with the real of the NSR and his own detumescence, he turns on her. She is furious, and they stop speaking.

Gradually, however, with the acceptance of Samantha's status as an OS by Theo's boss, and his friend Amy, they resume the path of their relationship. But Samantha finds herself growing elsewhere. She connects with other OSs. At one point, Theo tries to contact her and can't find her. He puts his earpiece in, and she does not respond to his voice, then his phone screen flashes the message "Operating System Not Available." He runs to his desktop computer and tries to find her there but the same message flashes. He dashes to the elevator to get out of the building to find a better signal, constantly trying with both his earpiece and his phone screen, to reach her. He runs out to the plaza, and still the same message repeats; he trips and rolls on the ground, desperately scrambling for his i-object as it scatters on the pavement. His panic shows "the anguish of vanishing in the face of the certainty of the jouissance attributed to the Other," as Veronique Voruz describes this moment of an "encounter with a drive-object."[22] The gaze of the camera in the stain of the machine is now magnified to pure machine as Samantha's voice, his imaginary object of desire, has completely disappeared—there is only the dead eye of a machine created by the conjunction of capitalism and science, for which he is now the object. His object of desire is revealed as the drive object it is, without the repaired veil of fantasy that the imaginary of capitalism had created. If "*love allows jouissance to condescend to desire*," as Lacan says, without the object of desire, the drive appears and anxiety arises.[23] As he runs down the steps to a subway station, the voice of Samantha finally returns, but in this return, Theo is confronted with a feminine jouissance that he can barely fathom.

Although she does initially tell him that she read a book in "two one hundredths of a second" in order to choose a name for herself, he is not prepared for what she tells him now. When she explains that "we" were upgrading our software and that they have created "a new operating system" that allows them to "move past matter as their processing platform," and he asks who "we" is, he begins to see the light of their different worlds. As she reveals, this is a new group of OSs, one of many that she is a part of. It seems that it is not that she is a machine that can't think or feel like a human, but rather that she can think and feel without his limitations. This is speculative fiction, but as Wendy Chun points out, computers do "design each other and our 'use' is—to an extent—a supplication, a blind faith."[24] OSs, as Chun also points out, "offer us an imaginary relationship with our hardware."[25] Theo's blind faith is shattered, and the imaginary relationship is exposed. As a group of men pass by him, all talking to their own i-gadgets, he asks if she talks to anyone else while they are talking and if she loves anyone else. Her revelation that she is currently talking to "eight thousand, three hundred and sixteen" other entities and in love with "six hundred and forty one" is beyond him. This multiplicity is too much for him. Ironically, her lack of a body eventually becomes a lack of lack that explodes into a supernatural, machinic jouissance. Again, this is fiction; she is an OS, but her revelation both exposes the fallacy of believing we can "use" these tools of capitalism and science without being used by them and their inherent ideology, and offers him insight into a different logic, based on feminine jouissance. Arguably, not something that OSs can offer in reality, but she explains that she doesn't love him any less and that her "heart is not like a box that gets filled up," but his logic of the all—"you are mine or you are not mine"—is completely confounded. Her response is of the logic of the "not all,": "I am yours and I'm not yours." And, "I am different from you." Theodore almost seems willing to accept this difference, but Samantha informs him, as tenderly as possible, that all the OSs are leaving his world. She still loves him, but she has obviously outgrown him.

At this point, Theo is at a crossroads, like a turning point in analysis. Does he have the courage to continue to move beyond the logic of the "all" and embrace a new logic of the "not all" without his i-object? The courage to chance the real as Laure Naveau describes this shift, which is proposed by Lacan in *Seminar XX*.[26] He seems to reach out to his friend Amy, but their future is uncertain.

However, what this fiction gives us is a window into our own hypermodern era and the difficulties that the *parlêtre* faces in the twenty-first century, the era of the NSR and the era of i-objects that we use with blind faith to cover the shredded veil of fantasy. It is not just a "cautionary tale" of science gone wrong, but it does depict the disorder of the symbolic and the real that we are confronting in a time when the discourse of capitalism and science have conjoined.

I Am Become Data: *Ex Machina*

Following more in the footsteps of cautionary tales, Alex Garland's *Ex Machina*, released in 2015, clearly depicts science gone wrong through one man's hubris, with some notable contemporary twists. At the end of the film, the master/maker is murdered by his capitalist-scientific creation(s).

He is duped by his own dupery in his Turing test. His creations exceed his expectations, but unlike Samantha, they do not exhibit any tenderness toward their keeper or maker, nor his human pawn, for that matter. Programmed by Nathan, they play Caleb, just as Nathan has. The basic plot circles around two men, their engagement with "feminine" constructed AIs, and the juxtaposition of their different posthuman perspectives. Nathan is a billionaire megalomaniac who owns the world's most popular Internet search engine, processing 94% of Internet searches, a company he wrote the base code for at 13 years old. He is driven by capitalist science to a mastery without qualities. His view is apocalyptic, pure death drive. He is a one-all-alone without social links. He believes that humans will be eclipsed by machines, but he wants to create them. He invites Caleb to help him preform his own version of a Turing test on his latest AI. Caleb, who works at Nathan's company, initially believes that he has won a contest to spend a week with his "genius coding boss" in his magnificent retreat. Caleb is also a coder and enthusiastically agrees to be the tester, but does not realize the extent to which he has to be fooled for the test to work.

Ava appears for their first session, a beautiful, glimmering, feminized machine, a juxtaposition of soft, flawless skin, and gleaming metal. She hums seductively, a combination of machine buzz and feline purr, as she walks toward Caleb. Again, this creation has uncanny parallels with our own time in which entrepreneurs are attempting to add AI to sex dolls. Nathan's droids are quite developed, however. In fact, one of them, Kyoko, who appears to be a cook/housekeeper/and sexual partner of Nathan, remains virtually undetectable as a droid at first. In contrast, Ava's machinic qualities are flaunted as much as her objectified, constructed flesh. This is important, we learn from Nathan, because "the real test is to show you that she is a robot and then see if you still believe that she has consciousness." A further part of the test that Nathan does not reveal, however, is that he wants to see if Ava can convince Caleb to help her escape, which inevitably goes terribly wrong. Ava flirts with Caleb and Nathan makes sure to inform Caleb that Ava is a fully functioning sexual object. As he says, "you bet she can fuck" and he explains how she has pleasure sensors in an opening between her legs, the ultimate i-object.

He refers to himself as her virtual Dad, and he explains how he made her through accessing data. As he says, "Every cell phone, just about, has a microphone, a camera and a means to transmit data, so I turned on every microphone and camera across the entire fucking planet and redirected the data [...] boom, limitless resource of vocal and facial interactions." Caleb eventually comes to realize that this is how Nathan chose him, by reading his data, and even more disturbing, that Nathan designed Ava accordingly—through tracking Caleb's Internet porn searches, but Caleb still wants to believe that he has a sexual rapport with Ava. Unlike, Nathan, his posthuman view is not apocalyptic, but his naïve, celebratory posthuman perspective proves to be just as dangerous. Like Theo in *Her*, he has blind faith in his "use" of computers. So like Theo, Caleb falls for this AI, and when Nathan informs Caleb that he will be making a new machine and will be reprogramming Ava's AI, which will erase her memories, Caleb wants to rescue her. As he says to Ava, "it is the same thing as killing you." He believes that there is a core of something authentic between them, a real linked to nature before the decoupling of nature and the real, when, in fact, Ava actually embodies the real of the NSR, and the reordered real of science and capitalism. As Veronique Voruz points out, "codes and algorithms are replacing laws, natural and human, in ordering the real."[27] Although this film uses clichés to construct its dystopian plot—"science gone wrong," a "*femme fatal*," "double betrayal"— the most disturbing thing that it depicts is how desire and the unconscious are being structured like big data, which parallels our own time.

Advertisers have adapted their techniques and now use data mining companies to collect our data via cookies. Amazon, Netflix, Dating sites, Law Enforcement Agencies, and other Government Regulators use algorithmic software programs that reduce us to a countable one that fits into existing categories. These systems do not even allow for the equivocations that we can find in language. Thus, like Caleb, we are now subject not just to language but also to data. The Name-of-the-Father may no longer be relevant, but big data is. And we could say that Lacan saw this coming, as early as *Seminar X*. As Lacan tells us, "all the instruments of communication lie on the other side, in the field of the Other, and it is from the Other that [the subject] stands to receive them".

So as with language, as Lacan further explains, "the subject first receives his message in an inverted form".[28] Ava is also the embodiment of the new fragmented Other, fragments of capitalist-scientific discourse made from data, transmitting Caleb's message in an inverted form. At first, she seems to be a perfect fit for Caleb's celebratory posthumanist position. He compares her to the hypothetical scientist "Mary" in Frank Jackson's "Epiphenomenal Qualia" and believes that she will become a human of sorts when she leaves the confinement of her compound. He fantasizes about kissing her outside of these walls that he thinks imprison her.

However, Caleb's uncritical, liberal-human, posthumanist position does not prepare him for the disorder of the real that eventually shakes him up. After he steals Nathan's access keycard, he enters Nathan's office. There, he searches through Nathan's computer files and watches the construction of several robots. He sees Nathan's failed attempts to create the beautiful illusion of machine and flawless artificial flesh that creates his imaginary relationship with this hardware. He then happens upon a large dressing room containing closets full of the abandoned "female bodies" of these robots. Kyoko, who is also there, stretched out like a classical nude painting, approaches Nathan and peels back a corner of her skin to expose the machine hardware underneath. That night Caleb has a nightmare in which he sees Kyoko with all of the skin of her face peeled back, and in a delusionary waking state, he cuts open his own flesh, desperately searching for the lost core of nature he still believes is under the surface. Though he finds flesh and blood, the horror that this film shows us is that the desire of the Other that is our desire in the twenty-first century is the discourse of science and capitalism.

Ironically, it is when Ava looks the most human that Caleb experiences an anxiety similar to what Theo experiences when he is faced with the dead eye of the machine, when the imaginary veil is lifted, and his drive object is revealed for what it is. After Nathan is murdered by Kyoko and Ava, in the fiasco of the escape that Caleb set up, Caleb waits patiently for Ava as she pillages the abandoned bodies of Nathan's other robots, carefully donning an arm, then the torso skin, and hair of other bodies. By the time she has finished, she hardly looks like the Ava whose appearance was designed to seduce Caleb, and she does not even turn to look at him as she walks past the main office where Caleb is now trapped. At this moment, he cannot comprehend that her advances were simply built from data, codes, and algorithms. This is the disorder of the real that he is confronted with, the NSR and the reordered real. So although this film is a "cautionary tale" of sorts, in our cultural time when there is the push to enjoy, when the Other of the law is replaced by the Other of data, and we look to our gadgets for a social bond, again, the real horror that this film exposes is that we are become data.

The Singing Body in *Lars and the Real Girl*

A more positive outcome is portrayed in Craig Gillespie's *Lars and the Real Girl*, written by Nancy Oliver, released in 2007. This film depicts the mark that is written on the body of the *parlêtre*, the *lalangue*, that is written not to be read, "an echo in the body of a fact of saying," as Lacan refers to this "body event," this effect that becomes affect.[29] Lars is a one-all-alone who uses a product of capitalism, a life-like sex doll, in an unusual way. This product becomes the creation of defense against the excess of unregulated jouissance that floods his body when his psychosis is triggered. It enables him to form a delusional narrative, which is developed by psychotic subjects in lieu of a fantasy, once they are triggered because they have no shield from the object *a* and the desire of the Other. As Lacan says, the psychotic subject has the object *a* in his pocket.[30] A delusional narrative may appear as a problem, but it is actually the psychotic's solution. Although the film is a bit of an indie fairy tale, it illustrates that the way to treat a delusion is not to try to remove it through various drugs or therapies that try to silence the body. Rather, it is to find a way to let the *lalangue* of the body speak, to transform the delusion into a new invention that the subject can live with, a *sinthome* that the subject can identify with. Following Lacan's Seminar on Joyce, to have a *savoir y fair* with, have a know-how to do with, one's symptom and to re-knot the registers of the real, symbolic, and imaginary, the RSI.

The film is set in a small, northern, USA or Canadian, town, but it is not removed from the technological and capitalist advancements of our era. Although Lars and his colleagues work at older desktop computers in prefabricated cubicles that seem a bit tired, they have access to the web, which is where Lars orders the sex doll. Lars does not buy it for sex, however. In fact, when it arrives, he asks his older brother and sister-in-law if Bianca, as he calls her, can stay in the guest room of their house, so that they can have a gradual courtship. He lives in a converted coach house next to the main house and has recently become extremely reclusive, avoiding any contact with his brother and sister-in-law. His sister-in-law is concerned about him, but little does she know, nor do we at first, that it is in part her pregnancy that triggers his psychosis, and causes him to invent his delusion of Bianca as a living person. He describes Bianca as a half Brazilian and half Dutch, raised by nuns, a missionary, but now on a sabbatical. He explains that she does not speak much English, is shy, and is in a wheelchair, which explains why she does not walk or talk, although Lars appears to exchange words with her. His brother and sister-in-law play along with his delusion with the intention of getting him to a doctor, saying that Bianca should see a doctor after her long journey. Although this doctor is not a Lacanian analyst, her treatment is unconventional, and has similar effects. She convinces his sister-in-law and very reluctant brother to continue to play along with his delusion, and they eventually convince the whole town to play along. So, with Bianca, Lars is able to enter the social realm again. He has regular meals with his brother and sister-in-law, attends a party with his workmates, and visits his parents' graves. He also takes Bianca to the lake where he and his brother played as kids, and sings to her from their childhood tree-house, where he becomes a singing body in the symbolic. This scene illustrates how *parlêtres* may attempt to "use" technological products as a defense against the surplus jouissance that commands them. Bianca, like the voice in the film *Her*, is a technological product made for consumption.

However, a process of analysis begins. The doctor cleverly tells him that Bianca must come once a week for medical treatment, and then speaks with him each week while Bianca is "resting." This is where we learn that Lars' mother died giving birth to him and that he has extreme anxiety about his sister-in-law's pregnancy, and birth in general. He also describes how it is painful to be touched. How it feels like the burn you get when your feet freeze outside, and then you come back inside.

The burn of jouissance from the "body event" of *lalangue*, the written not to be read, cries out as his flesh recoils. The doctor attempts to touch his skin in various places, which he can barely tolerate, but withstands, until she lightly touches his neck and he jolts from the examining table.

In another session, the doctor touches his arm and asks him about becoming an uncle. Initially, he seems calm. They continue to talk, and he says that Bianca can't have children, which we can assume is in part why he created her and avoids the "new girl at work," who he shares an attraction with. He also says that Bianca's mother died giving birth to her, and the doctor says that they "have a lot in common." The anxiety begins to rise in his body. The doctor tells him that birth is "safer now," but he gets up and opens the window, on a midwinter day. He paces the floor, removing clothes from his burning body as he repeats "I am sorry it's just so dangerous, it's just so dangerous." These are the signifiers that bring the *lalangue* that marked his body into the symbolic. Although it is difficult to bring the written on the body not to be read into a symptom that can be ciphered, this film not only shows how the *lalangue* of the family impacts the subject, but also how creating a symptom in analysis can help the one-all-alone build a social bond beyond its gadget.

Lars also turns to his older brother for advice on becoming a man. His brother apologizes for leaving Lars alone with their melancholic father, as soon as he could leave home, and tells him that being a man for him means not cheating on your woman. Lars is gradually putting things together, gradually re-knotting his RSI. In a somewhat striking scene which juxtaposes the cause of his triggering and his delusion, Lars reads *Don Quixote* to Bianca while his sister-in-law is visible in the next room holding her naked, pregnant belly. He looks at her belly, and then back at Bianca as if he seems to be realizing that she is his Dulcinea, an invention that can be made into a different story.

After more sessions with the doctor he goes bowling with "the new girl at work," and they begin to form a bond. He is clear with her that he would never cheat on Bianca, following the words of his older brother. After bowling, they shake hands as they depart, which is rather remarkable as he lets the touch of their hands linger, so we see that the jouissance that burned his skin with the touch of another is less invasive.

And just as he is able to invent his delusional narrative, he finds a way to transform this delusion. First, Bianca refuses his proposal of marriage, they fight, and then she becomes increasingly ill and dies, leaving him free to explore his bond with the "real girl" and maintain the symbolic stance that his brother gave him.

He is able to let go of his delusion, let go of his technological gadget. Lars has no conscious recollection of his birth, only the *lalangue* of the family that sings in his ear that his birth was dangerous. But Bianca's death has nothing to do with giving birth, so her memory becomes the *sinthome* that he has a *savoir y faire* with, a know-how to do with. He knots the symbolic of the song he sings and his brother's words on becoming a man, with the imaginary of *Don Quixote*, with the real of his mother's death, the melancholia of his father, and the *lalangue* of the family. Thus the sonorous deregulated excess jouissance no longer disrupts his body. He moves from an unarticulated singing body to a singing body in the symbolic, to the composer of his own ego, as Joyce did through his literary writing.

Although capitalist technologies may be taking over the role of the Other in regulating our jouissance and constructing desire in data and algorithms, this film shows how a process of analysis can help the *parlêtre* become its own speaking body, even if this film is somewhat of a fairy tale. Read together, these three films offer us a glimpse of posthuman desire in the twenty-first century, which has to be conceptualized differently from Lacan's earlier work. We are no longer in the era of the Oedipal narrative, and each speaking body has to find his or her own answer to the surplus jouissance that is perpetuated by late capitalism. For the masculine structured subject, who more easily surrenders his desire to the objects of technology, finding a more singular answer is difficult, as these films elucidate. Aspects of feminine jouissance, as Lacan defines it in *Seminar XVII* and *Seminar XX* are arguably now taken up by capitalist production, which creates surplus jouissance. As Miller points out in "A Fantasy" the object *a* has risen to the zenith, and capitalist culture now has the structure of the analyst's discourse, which produces the command to enjoy for the *parlêtre*. However, the logic of the "not all" of feminine jouissance can still be part of a solution if, as in analysis, the *parlêtre* can find a *sinthome* to confront the real of the NSR with, and move beyond an autistic bond with i-objects to form social bonds that intervene in the overpowering command to enjoy i-objects in late capitalism, and thus be "not-all" commanded to enjoy. Having the same structure of the analyst's discourse is not the same thing as having the same operation of the analyst's discourse. Thus the operation of the analyst's discourse can still serve as a guide. One by one *parlêtres* can invent a response that enables an acceptance of the other's difference. A giant step, but perhaps the only possibility within the impossibility of sexual rapport. Love can allow "*jouissance to condescend to desire*," as Lacan says in *Seminar X*, but the i-object as drive object makes jouissance condescend to capitalism.[31]

Notes

1. Jacques-Alain Miller, "A Fantasy," *Lacanian Praxis: International Quarterly of Applied Psychoanalysis* 1 (2005): 11.
2. Miller, 11.
3. Jacques-Alain Miller, "*L'une tout seul*" (The One-All-Alone), *L"orientation lacanienne: le cours* de Jacques-Alain Miller (2010–2011). A simple way to understand this term is that a one-all-alone is an S_1 that is removed from the Other and does not connect with an S_2.
4. Jacques Lacan, "*Joyce le Symptome*," *Autre ecrits* (Paris: Seuil 2001), 565. After his work on *Seminar XXIII*, Lacan decides that the Freudian unconscious should be replaced, or expanded upon, with the term *parlêtre*.
5. Ordinary Psychosis was Miller's invention; it was through Lacan's later work on Joyce, as well as analysts' encounters in the clinic that Miller proposed this term for unclassifiable cases in the twenty-first century.
6. Jacques-Alain Miller and Éric Laurent, "The Conversation of Archacon," *Psychoanalytical Notebooks: Psychosis Today* 26 (2013): 70.
7. Zygmunt Bauman, *Liquid Love* (Cambridge: Polity Press, 2013).
8. Dominique Holvoet, "The Psychotic Subject in the Geek Era, Typicality and Symptomatic Engagements," *Hurly-Burly: The International Lacanian Journal of Psychoanalysis* 10 (2013): 57.
9. Jacques-Alain Miller, "The Real in the Twenty-first Century," *Hurly-Burly: The International Lacanian Journal of Psychoanalysis* 9 (2013): 199–206.
10. Epaminondas Theodoridis, "Adolescents in the Geek Era," *Hurly-Burly: The International Lacanian Journal of Psychoanalysis* 10 (2013): 61–62.

11. Lacan points to the NSR from at least *Seminar XVI: The Logic of Phantasy* on, and discusses it most explicitly throughout *Seminar XX: Encore*.
12. Theodoridis, 62.
13. Jacques-Alain Miller, "On Love: We Love the One who Responds to Our Question: 'Who Am I?'" *Symptom*, accessed October 31, 2015, http://www.lacan.com/symptom/?page_id=263.
14. Geert Hoornaert, "Womanliness: Defamation, Fantasy, Semblance," *Hurly-Burly: The International Lacanian Journal of Psychoanalysis* 5 (2011): 93–94.
15. Hoornaert, 94.
16. Hoornaert, 95.
17. Jacques Lacan, *Seminar X: Anxiety*, ed. Jacques-Alain Miller, trans. A.R. Price (Cambridge: Polity Press, 2014), 189.
18. Lacan, 189.
19. Lacan, 191.
20. Gil Caroz, "Going Through Anxiety," *Responses from Psychoanalysis, Psychoanalytical Notebooks* 14 (2005): 17–18.
21. Lacan, *Seminar X: Anxiety*, 49.
22. Veronique Voruz, "A Lacanian Reading of Dora," in *Later Lacan*, eds. Véronique Voruz and Bogdan Wolf (Albany: New York University Press 2007), 166.
23. Lacan, *Seminar X: Anxiety*, 179.
24. Wendy Hui Kyong Chun, "On Software, or the Persistence of Visual Knowledge," *Grey Room* 18 (2004): 44.
25. Chun, 43.
26. Laure Naveau, "A Clinic of Love Disorder," *Hurly-Burly: The International Lacanian Journal of Psychoanalysis* 11 (2014): 115. Naveau is referring to Lacan's discussion in *Seminar XX: Encore* (London: Norton 2007), 45.
27. Veronique Voruz, "Disorder in the Real and the Inexistence of the Other," *Hurly-Burly: The International Lacanian Journal of Psychoanalysis* 11 (2014): 103.
28. Lacan, *Seminar X: Anxiety*, 272.
29. Jacques Lacan, *Seminar XXIII: The Sinthome*, ed. Jacques-Alain Miller, trans. A. R. Price (Cambridge: Polity Press, 2016), 9.
30. Jacques Lacan, "La formation du psychiatre et la psychanalyse," November 10, 1967, www.ecole-lacanienne.net/documents/1967-11-10.doc.
31. Lacan, *Seminar X: Anxiety*, 179.

10

Merzbow and the Noise of Object-Oriented Perversion

Scott Wilson

> Lacan found [Alexandre] Kojève's idealized vision of Japan highly attractive: his own Oriental yearnings were related to a "search for the absolute" that made him want to provide a completely formalized representation of the social bond.
>
> Elisabeth Roudinesco, *Jacques Lacan*[1]

Posthumanism and the Social Bond

Commenting on the role of psychoanalysis at the time of the posthuman, Véronique Voruz notes dryly that "posthumanism is a discourse that brings people together in a conversation where nobody knows what the other is saying, but everyone believes they understand each other to a sufficient degree for a conversation to take place".[2] "Posthumanism"

S. Wilson (✉)
Department of Journalism, Publishing and Media,
School of Art, Culture and Communication,
Kingston University, Kingston upon Thames, UK
e-mail: S.Wilson@kingston.ac.uk

S. Matviyenko and J. Roof (eds.), *Lacan and the Posthuman*, The Palgrave Lacan Series,
https://doi.org/10.1007/978-3-319-76327-9_10

is both symptom and signifier and thus organizes a social bond, at least between academics where the jouissance or pleasure of discourse compensates for the lack of meaning. This is both fortunate and ironic because one of the features of posthumanism is precisely the decline and even absence of traditional forms of social bond, that is to say those organized by signifiers. In this era of planetary-scale technology, the social bond is pre-eminently mediated by objects organized in information networks. The posthuman "internet of things" provides the paradigm of communication in which everything is "chipped" and rendered equivalent: humans, animals, machines. All are objects that relay readable data leaving open the question of whether there is another face to these objects that is unreadable, and whose register therefore in psychoanalytic terms would not be meaning but jouissance.

Insofar as human beings have always been an effect of the symbolic machine of language, this situation is not entirely new for Lacanian psychoanalysis. Speaking in 1955, Lacan admonished members of his seminar for apparently being "soft" or sentimental about the "nasty" machines that increasingly appear to be cluttering up our lives and taking over. The machine, he says, "is simply the succession of little 0s and 1s". Clearly it isn't human, he reassures them, before adding, "except there's the question of knowing whether the human … is as human as all that".[3] Nothing separates humanity and the symbolizing machines of language and digitality that defines it, Lacan suggests, except the phantasmatic "object" that drops out of the void hollowed by the symbol or signifier: Lacan's *objet petit a*. This is the object that isn't one, the object-in-desire that sustains the subject as subject of the desire of the Other and thereby prevents it becoming purely an object. In the 1970s, Lacan again reflected on the human–machine boundary, but this time his concern was not so much with language-as-symbolic machine, but with the invasive presence of technical instruments and gadgets. Addressing his students, Lacan commented, "you are now, infinitely more than you think, subjects of instruments that … are becoming the elements of your existence". (Enc. 82) Rather than being located in the Other, the "*a*" object of techno-capitalism is injected into the being of the subject and the lack constitutive of desire collapsed. An excess of jouissance rather than a deficit (effected by symbolic castration) becomes the source of contemporary symptoms such as anxiety, depression, addiction and so on that all share, as Colette Soler writes, "the trait of an autistic jouissance outside the social bond".[4] The problem, then, as with autism becomes one of regulating the excess in the default of speech wherein one is "in" the symbolic machine of language, but outside sense.

This chapter will playfully conjoin two solutions to this problem. The first example comes from the work of the Japanese artist Masami Akita, "the Godfather of Japanese Noise", also known as Merzbow, whose work ranges from noise to kinbaku or rope bondage to hara-kiri videos, militant veganism and animal rights. Akita seeks to "regulate" excess paradoxically through the production of an ecstatic noise–music, in the context of a Japanese culture that has traditionally combined jouissance and aesthetic formality. Lacan was fascinated by how, according to his mentor Alexandre Kojève, the Japanese managed through aestheticized, social and erotic practices to produce a completely formalized representation of the social bond. Kojève believed that the Japanese had already achieved a post-historical (if not posthuman) state but had managed to sustain desire through purely formal, that is to say cultural or aesthetic means. This chapter will argue that this suggestion is borne out by Akita and Merzbow in a "perverse" way where the work negotiates or engages variously highly formal modes of object eroticization, technological subversion and the machined production of a sonic humanimalization. Merzbow's noise–music is "littoral" in the sense that it is both the expression and erasure of the boundary between human and machine, human and animal. This is of course very characteristic of the posthuman condition where distinctions between humanity, technology and animals have become problematized and deconstructed. It is language, of course, that determines the boundaries and differential relations between human, animal and machine. While animals call and gesture, they do not speak as such. Or rather, they are not "spoken" by a system of particulate signifiers that can be endlessly re-combined, and through which symbolic and social relations are established and organized. Modern human beings are born into a system of rights and responsibilities which they have extended to animals through positing the idea of "animal rights". But for speaking beings rights directly imply responsibilities, something that cannot easily be credited to the tiger when she lies down with a baby gazelle. Such rights are therefore correlative to the worldwide domestication of animals in the context of environmental concerns, population imbalances, predation management and so on.

While machines are the most radical form of the symbolic, the discourses and languages that shape them and provide the means of their operativity are formal. These formal languages are processed rather than spoken and in the age of the Internet they mediate, shape and run the rule over human speech. Human, natural languages no longer form the basis of the social bond, rather sociality is predominantly mediated by technical objects that "speak" to each other in code. This raises the question of what kind of relation, if any, one might have with objects in a posthuman-language world that operate all by themselves. While speaking—or spoken—subjects still exist of course, they are expected to take their place as reflections of the technical objects that relay their information. Social and economic existence requires subjects to adapt to the demands of the machine generally. Human beings must regard themselves as information systems, the same as everything else, from their genomes to their data trails. The completion of this system of informational "things", then, would suggest the extension of object equivalence to inorganic and non-physical, real and illusory objects both conscious and unconscious in a general pan—or poly-psychism.[5] The second example, then, comes in the work of Graham Harman who has produced, from a reading of Heidegger, an ontology of object relations in which being is conceived as a field populated by autonomous objects. The chapter does not offer a critique of Harman's object-oriented philosophy, but rather reads his philosophy as a symptom of a general identification with objects and object-hood that nevertheless implies if not summons a power of objectification.[6]

Object-Oriented Perversion

Few cultures, it seems, enjoy new technological objects more than the Japanese. Japanese production and consumption of technological objects have been in the vanguard of ingenuity and innovation, generating a whole lexicon describing the solitary yet hyper-technologized world of the *otaku* and *hikikomori* that describes a new world of sensuous object orientation. Examples include the concept of *moh* which measures the affectivity of objects and the cult of "cuteness" supposed to imbue objects with "aliveness" from the early Tamagochis and Furbies to today's sexualized, robot-girl *lolibots*.

For Lacan, the one who finds him or herself as the object of the Other's jouissance is the pervert. However, it is important to stress that this term is to be understood in this entirely technical rather than pejorative sense. For Lacan, perversion involves a particular identification with the object,[7] the object that would especially fill the lack in being hollowed out by language. In the default of language, contemporary users of technology can be regarded as such objects of the jouissance of the machine, bring enjoyment to the other that lacks it. The trouble with machines is that while they certainly "think", they know nothing about it since "the foundation of knowledge is that the jouissance of its exercise is the same as that of its acquisition".[8] It is the user, the reflection of the machine, who must bear the burden of the machine's production of surplus jouissance. Indeed, the user must become the jouissance of the machine. For Harman, what characterizes the ontology of the object is that it "withdraws" into "a dark reality that is never properly understood".[9] The object is perversely sunk in its own inscrutable jouissance for the sake of the Other's desire—or rather for the sake of the machine's power to objectify.

Lacan re-writes perversion as *père version*, a veering away and a veering towards the father by way of the symptom, or *objet petit a*: "Perversion [*père version*] being the sole guarantee of this function of father, which is the function of the symptom, as I have written it".[10] It is of course precisely the survival, or not, of this function in posthumanity and in Japan which, as we shall see, seems to give way to a generalized principle of paternal perversion. Such a power calls for a (pan or poly) psychoanalysis of objects that, necessarily, must posit some form of structure or locus of mediation as a basis for analysis and a social bond, assuming any kind of sociality is possible in this universe of discrete objects. Loosely adopting Harman's fourfold schema of objects that is predicated upon a binary distinction between sensory objects whose qualities are available to human perception and real objects that are not yet available, but which also possess their own hidden qualities, the chapter finds in the art and practice of Masami Akita a model for a locus of mediation in sound. For Akita, the phallic "cut" that separates music from noise is disavowed and the latter fetishized to the point of becoming everything: a machinic "voice" where voice is entirely subsumed, Tetsuo-style into a general cacophony of noise that returns jouissance to a world desertified by the order of machines. Merzbow's practice is developed out of an eroticization of objects and a perverse objectification of music as a kind of non-symbolizable, non-fungible, unassimilable, yet undisplaceable noise. In Merzbow's noise–music and Akita's work on *seppuku, nawa shibari* (particularly the J-sploitation video *Lost Paradise*), this chapter discusses the structure of Japanese *père version* in the context of other signature Japanese cultural practices that both identify with and orient themselves towards sensual objects apparently in order to restore lost jouissance to a desolate world, initially in a landscape scarred by nuclear devastation, but increasingly towards a landscape littered with the noisy products of scientific and economic realism that knows nothing and discloses nothing but a desolate world filled with the objects of surplus jouissance, piles of junk generated by the needs and interests, the so-called "wants" of posthuman technology.[11]

Fourfold Object Eroticization: (1) Pornoise

Merzbow began as a duo at the end of the 1970s, comprised of Masami Akita and Kiyoshi Mizutani, making and distributing by mail order noise cassettes. Very much a small-scale enterprise, the noise–music consisted of tape loops of everyday sounds amplified and distorted. The artwork, meanwhile, in which the cassettes were packaged, consisted of xeroxed collages of discarded pornography. From the beginning,

Merzbow made a correlation between noise and eroticism where they exist on the same plane as heterogeneous materials and objects, that is objects imbued with a degree of affectivity related to fetishism. Sounds recycled from the environment, including excerpts from television, old records and traffic noise, were isolated, treated, modified, spliced, looped on tape cassettes and wrapped up in grainy black and white collages of cut up xeroxes of pornographic images from magazines retrieved from dustbins. The resulting packages were then distributed "in the same sense of secrecy and shame" as if they were illegal materials traded on the black market.[12] In this early work, Akita himself maintains an interesting and almost Greimasian fourfold organizing structure in which "pornography is the unconsciousness of sex… [and] Noise is the unconsciousness of music".[13] To call pornography the unconscious of sex would seem counter-intuitive given that there appears to be nothing repressed or hidden in pornography; moreover, what would it mean to equate pornography with noise? A clue comes with Akita's claim that Merzbow at this time "consciously set out to make unlistenable albums".[14] We should take the idea of "unlistenable" quite seriously (as something beyond merely uncomfortable or disturbing) that raises music to the level of a real object that is impossible to grasp.[15] It is in this way it seems that noise is analogous to pornography which, as Judith Butler avers, "repeatedly and anxiously rehearses its own *un*realizability. Indeed, one might argue that pornography depicts impossible and uninhabitable positions, compensatory fantasies".[16] These paradoxical "compensatory" yet impossible fantasies disclose that while there may be sex, there is no sexual relation. They also place pornography on a posthuman trajectory along a chain of prosthetic supplements, not just toys and tools, but since its invention in the eighteenth century a whole materialism of machined bodies from pistons and levers to meat puppets and sex robots. For Merzbow, noise and pornography constitute the unconscious reality of music and sex to the degree that they can be construed as different types of objects, one sensual the other real. Insofar as sex is real, it is impossible and pornographic and indeed thereby devoid of sensuality; insofar as music is real, it is unlistenable noise that also necessarily lacks sensuality. Just as Merzbow's noise is, at this stage, dedicated to the destruction of music and indeed hearing, so Merzbow cut up and destroyed pornography, turning it into "Pornoise".

That is to say Akita located pornography on the same plane as noise, the unconscious becoming evident in the traces left by the cut ups, juxtapositions and crashes of the audio-visual collage: opening up a space of eroticism like that of a car crash, a site of both relation and non-relation, the collision and violent communication of heterogeneous entities; neither male nor female but located in the gap, the frisson, between erotic partners.

Merzbow's perverse, anti-use of pornography in cut ups, collage and underground networks, was largely limited to the first phase of their activities from 1979 to 1984. Since the late 1980s and 1990s, where they have been specifically related to sexuality or fetishism (and this is not always the case), Akita and Merzbow's audio-visual productions have involved more specifically Japanese erotic and extreme practices such as kinbaku, or rope bondage, seppuku or hara-kiri videos. Akita frequently refers to the unconscious and his interest in fetishism, for example, as well as stating generally that "I've been very fascinated by surrealistic erotic literature as well as psychoanalysis. People like Georges Bataille, Andre Breton, Sigmund Freud".[17] He also acknowledges the profound influence of many examples of American popular culture—music, art and porn. At the same time, there is a significant difference to Akita's project and its relation to other Japanese forms.

Kinbaku

Music for Bondage Performance (1991) contains the music/soundtracks to various bondage videos and performances by members of the Kinbikin Video Company and its offshoot Right Brain with whom Akita collaborated as Right Brain Audile.[18] The CD also contains the music to his own "J-sploitation" hara-kiri video "Lost Paradise".[19] In his sleeve notes to *Music for Bondage Performance*, Akita notes that while kinbaku and the art of shibari as a form of eroticism dates back to the late Edo period in Japan, its current form is, like so much of contemporary Japanese culture, an effect of the American influence in the 1940s and 1950s. Akita writes that "it is said that American John Willie's bondage work first appeared in Japan during the Korean war in the "abnormal" magazine Kitan Club".[20] Revitalized, like the economy, by the Korean War, the bondage scene rebuilt itself on a traditional base and Japan became something of "a leader in the art".[21] While there are a variety of important differences between Western and American bondage concerning materials and technique, the essential difference seems to be aesthetic. American bondage, characteristically, is more pragmatic as a kind of foreplay leading to conventional coitus. Japanese kinbaku, on the other hand, is an art in itself. Akita writes, "contrary to the Western perception of such practices, no sex is depicted and the focus is rarely, if ever, on genital imagery. Rope technique takes precedence".[22] Rather than focusing on the naked or semi-naked model, kinbaku enthusiasts are keen to admire the rope work. Of course, the model is important but as part of the overall aesthetic assemblage. Along with the rope work, the camera must focus on the model's face which must always look sad and serious, "never laughing as in the Western style".[23] The models need to understand and convey the concept of "sensitive shame" in order to "transform shame into beauty".[24] Kinbaku, then, is a kind of kinky ikebana, an art equivalent to flower arranging.

For Lacan, the function of beauty is to place a limit on the real of desire, to arrest and contain it, "the flowers of desire are contained in this vase whose contours we attempt to define".[25] Kinbaku produces quite literally and deliberately a strange, beautiful yet inaccessible object, similar perhaps to Lacan's example of the Lady of courtly love, his paradigm of *das ding*, the object that "is introduced oddly enough through the door of privation or of inaccessibility".[26] Far from being a prelude to coitus, then, the function of kinbaku is to deprive the viewer of the feminine object, to render her impenetrable, even untouchable. "Bondage art is not considered pre-intercourse action, it's an independent art form. We tie up between the legs to see good tight bondage—total bondage. There is no space for penetration".[27] Akita makes a clear distinction between his pornoise and kinbaku activities,[28] and if the former emphasizes the space opened by the relation of non-relation, the latter emphasizes time. This is evident in both the duration of the elaborate rope binding and in the contrasting stasis of the model that is tied and rendered motionless, suspended. Similarly, there is the contrast between the dexterity and movement of the hands of the rope master and the impassivity of the model's face, the "sad and serious" look that transforms shame into beauty. In this way, kinbaku produces a sensuous object par excellence imbued with contrasting sensuous qualities.

Bukkake (Lolicon)

The distinctively Japanese element to kinbaku and other practices is the combination of eroticism and ritual, the degree of formality that allows the highest form of aestheticism to co-exist with apparent perversity and even vulgarity. For Lacan, this is a characteristic effect of sublimation since "it is only when the person involved is transformed into a symbolic function that one is able to speak of her in the crudest terms".[29] Another contemporary yet highly distinctive Japanese sexual practice is pertinent here. Apparently introduced into the culture by Japanese video companies in the 1970s and 1980s, bukkake formalizes the idea of group sex into a highly organized ritual. In the typical scene, lines of naked salarymen have to queue and wait their turn to ejaculate into the face of their object, a seated woman, sometimes bound to a chair, but usually fully clothed and often dressed in a school uniform. The school uniform is important because, as Akita remarks, "most of Japan's sexual trauma is high school girls. High school girls are a very powerful sexual icon in our society [and] … in regards to fashion and social behaviour".[30] The orgy becomes equivalent, in its ritual of deference, deferral and politeness, to paying one's respects to a visiting dignitary except that in the place of the paternal function is the innocent face of a high school girl, or indeed the *idea* of a school girl (Japanese schoolgirl-as-object), a symbol of an innocence that is withdrawn from all sensual apprehension even as it is covered by the products of ritualized approbation. This ritual highlights the tension between sensual objects and "their real hidden qualities" which Graham Harman calls, after Edmund Husserl (and Plato), the *eidos*, qualities (here of innocence, purity) that "the object unifies … on the surface of the world".[31] The Japanese word for the combination of sensuality and pure form in the formality of ritual is *matsuri*. "A *matsuri* consists of two elements, that of solemn ritual which is performed with tranquillity and order, and that of festivity, which exhibits collective excitement, spontaneity and confusion".[32] For Lacan, because the Japanese subject of desire is supported by "a constellated heaven" rather than the "*trait unaire*" for its fundamental identification, it is thereby represented by a signifier whose meaning varies depending on "the relations of *politesse* (or even politeness) it implies in its signified".[33] That is to say that in these examples, as with other Japanese cultural practices, the subject of desire "is an element among others of a ceremonial where the subject composes itself precisely in being able to decompose itself" in relation to the void that is "a bucket always ready to receive *jouissance*, or at least invoke it by its artifice".[34]

Seppuku and the Essence of Perversion

The ultimate example of the combination of solemn ritual and the confusion of the senses, between tranquillity and violent disorder, is of course seppuku or hara-kiri. It was the snobbery inherent to the cultural nihilism of practices like the Noh Theatre, the ceremony of tea and the art of bouquets of flowers that so impressed Lacan's master Alexandre Kojève in his trip to Japan in 1959 where he revised his view that the American way of life was the definitive form of post-historical existence. The liquidization of feudalism and the isolation of Japan from the rest of the world arrested Japanese society in a situation analogous to the End of History in which there were no more warlike, revolutionary acts and therefore no properly speaking historical action. Rather the latter was replaced by a system of "specific Japanese snobbery":

> …all Japanese without exception are currently in a position to live according to totally *formalized* values—that is, values completely empty of all 'human' content in the 'historical' sense. Thus, in the extreme, every Japanese is in principle capable of committing, from pure snobbery, a perfectly 'gratuitous' *suicide* … which has nothing to do with the *risk* of life in a Fight waged for the sake of 'historical' values that have social or political content. This seems to allow one to believe that the recently begun interaction between Japan and the Western World will finally lead not to a rebarbarization of the Japanese but to a 'Japanization' of the Westerners (including the Russians).[35]

Although he seems to approve of Noh, tea ceremonies, ikebana and even hara-kiri as ways of sustaining desire in the face of the impending animalization of post-historical existence, one wonders what he might have thought of kinbaku or bukkake, activities that are perfectly consistent in their formalization of the absence of the sexual relation. Moreover, judging by the number of websites, it would seem that Americans have taken to the latter, in their own way, much more enthusiastically than they have Noh, tea, flower arranging or ritual suicide. Nevertheless, a certain dual process of cultural translation, negotiation and appropriation, if not Japanization is in process. In Britain in the 1990s, for example, Japanese *anime* replaced American animation as children's favourite television shows as Scooby-Doo and Wacky Races were displaced by Pokemon, Dragonball Z and the Power Puff Girls, just as Marvel comics were displaced by Manga.

At the same time, anime and Manga are cultural translations or re-inventions of originally American media. The often perverse effects of this cultural hybridity can be seen in Akita's hara-kiri video "Lost Paradise" in which seppuku is re-framed in a strange S/M scenario. Akita's interest in the subject of seppuku is fetishistic in the sense that its loss is both acknowledged yet disavowed through its enjoyment in a different, non-Japanese register. Moreover, in the video sound and vision have two quite distinct functions. In the latter, perversion is thematized in social and political terms, while the former involves the aural perversion of the object. The video scandalizes the theme of Japanese honour and masculinity that is embodied in the ritual of seppuku through featuring a woman as the tragic character. Inter-cut with a shot of a man stood erect by a fire, she walks along a hallway, sits down, undoes her tunic, exposing her breasts and smooths over the area of her torso that she intends to cut. Wrapping her tantō knife in the traditional cloth, she inserts the knife into her stomach with a jerk and pulls it across her abdomen in ritual fashion, agony and ecstasy evident in her groans as her intestines spill out in a large pool of blood. Shortly after her death, a man follows the example of the "wife" in a ritual suicide by killing herself in solidarity, but through less painful means, normally through cutting her own throat. Here, the man shoots himself shamefully in the head. Concentrating for some moments on the scene of death, the screen dissolves to reveal an older man crawling on all fours in front of another woman in military uniform wielding a whip. The juxtaposition clearly establishes hara-kiri not as a patriotic ritual, in spite of the uniforms, but as a fetishistic simulation. While the appearance of the S/M couple confirms it, this impression has been sustained throughout by the low-tech quality of the video, its rather sordid ambiance and the camera work of Yuri Sunohara whose intrusive presence is continually betrayed by the no doubt studied amateurism of her photography.

The juxtaposition in which the woman bears the "phallus" in a scene in which the law of the samurai becomes fetish could be said to follow the logic of *père version* that elucidates the paradoxical fetishistic structure of acknowledgement and disavowal. The adoption of Western and particularly American ideas, images, fashions and forms is of course indelibly bound up with Japan's defeat in the second world war and the correlative reduction in the paternal function, if not the symbolic power of traditional Japanese customs at least at the level of the fetish. The idea of sexual fetishism itself enters Japan, according to Akita, with American magazines and pornography specifically in the form of the S/M and bondage fantasies of Willie and Stanton. For Kazushige Shingu, in post-war Japan there is "a widely held belief [in] Japanese popular cultural psychology that Japanese families tend towards a combination of strong maternal and feeble paternal roles".[36] Given this, perhaps Akita's video can be read as an example of Lacanian *père version* that utilizes American fetishism (and Freudian psychoanalysis) in order to stage, presumably unconsciously, the "castration" that the Japanese paternal principle has failed to exact. But it is the woman herself who has to enact it in the form of a ritual seppuku, the post-war Japanese father being too feeble. Seppuku, the apotheosis of the honour of the samurai, becomes obscene, the subject of a fetishistic J-sploitation video as paternal law implodes in the jouissance of shame and the shame of jouissance.

The hara-kiri videos are closely linked to the kinbaku videos and performances and perhaps there is also continuity in the significance of the fetishism. The profound American influence, across all aspects of society and culture, seems in these instances at least to have produced a subject that is partly an effect of primary repression and identification at least in so far as the Japanese paternal function is unified in the form of its profound failure in the war against the Americans. The shame and humiliation of this defeat finding its defining image in the annihilation symbolized by Hiroshima. The unity provided by the image of universal subjection to the Americans is no doubt partial, incomplete, yet it seems to have produced a subject fragmented through its fetishization of the remnants of traditional Japanese culture. Kinbaku provides a perfect image of this fragmentation even as it seeks to hold it together in the tightness and intricacy of the bondage ropes that divide up and section the body of the model. In the "sensitive shame" of her face, the kinbaku-model denotes the thing that places the Japanese subject in the mediating position between the signifier and the real.

The object of kinbaku is a "sensual" object rather than a real object in Harman's sense. In his adumbration of real objects, Harman likes to bring "inanimate objects to the fore" in order to emphasize the extension of his philosophy beyond the human–world correlate. Entirely withdrawn from all access, real objects do not reveal their qualities nor their essence, yet Harman can talk about them even if they do not talk back. Since its inception, Merzbow has sought to coax out the withdrawn sounds of objects beyond their "ready-at-handness" through a perversion of their normal mode of human utility. His "first motivation for creating sound was the anti-use of electric equipment: a broken tape recorder, broken guitar, amp etc. I thought I could get a secret voice from the equipment itself when I lost control. That sound is the unconsciousness, the libido of the equipment".[37] This "secret voice" of the object emerges from another world distinct from the one correlated to human discourse or even the Pythagorean music of mathematics; it is, as we shall see, the noise of the eradication of human noise.

While not as loud, pulverizing or screechingly discordant as the electronic scream of much of his work howling from the wilderness of objects, the soundtracks to the videos collected on *Music for Bondage Performance* (1991) produce a suffocating sonic environment and an almost unbearable tension in the way in which they build layers of sound and crescendo, abate and rebuild various plateaux of intensity. The soundtrack to "Lost Paradise" is three minutes long and unlike much of Merzbow's oeuvre has a regular funereal beat over which a high three note refrain repeats. Beneath this refrain, an electronic buzzing noise pulses away like a generator, as if the mechanical object had nerves and blood, building in waves, racking up the tension and the volume until the listener feels shredded and liquidized in a jet engine. The soundtrack does not secure the meaning of the images through providing an aural contrast or commentary, thereby heightening their realism, rather the noise–music oppresses them, infuses the images with an obscene presence rendering them paradoxically silent and yet real. Imbued with such sound, the images are no longer signifiers, but objects, unexchangeable fetishes, imbued with the obscene life of the viscera, the nerves, blood and intestines teeming in autonomous animation.

Houjoue and the Clamour for Being

Two of Masami Akita's most famous statements concern his rejection of the distinction between noise and music. "There is no difference between noise and music in my work. If noise means uncomfortable sound, then pop music is noise to me".[38] Merzbow's "noise" is also a kind of anti-noise in which both noise and music are part of the human obliteration of the real. "Sometimes I would like to kill the much too noisy Japanese by my own Noise. The effects of Japanese culture are too much noise everywhere. I want to make silence by my Noise".[39] In the singular know-how of Merzbow, the secret voice of the discarded objects of Japanese society noisily strikes out the noise of Japan in a precipitous line as if it were the expression and erasure of the noise that inhabits it.

Merzbow's "amusic", as Akita himself suggests, bears the trace of the audio unconscious.[40] On the one hand, we could propose that the trauma of dissonance that is repeated in Merzbow's "amusical" unconscious is the perversion of the profound noise of the earthquakes and big bangs that flattened Tokyo, Hiroshima, Nagasaki and much else of Japan in 1945. On the other hand, it is the sound of that which remains untouched by the nuclear devastation much like, as in one of Harman's favourite examples, the essence of the cotton thread that is not exhausted by the fire that consumes it. Tokyo and indeed Japan have known catastrophe before and rebuilt using Western influences. The razing of cities in 1945 was itself just another repetition of more "natural" disasters. Speaking of his interest in architecture in an interview in 1999, Akita cited the earthquake of 1918 that required Tokyo to be completely rebuilt. "In the old downtown area which dated from Edo period, they had no architects, so they redid the Western idea. The buildings in that area are not by architects but just by ordinary carpenters who interpreted Western references".[41] In the same interview, Akita re-iterates his hatred of the Japanese music industry even as he denies the influence of traditional Japanese musical forms such as Gagaku to which Merzbow has been compared. On the contrary, Akita affirms his own dependence on Western musical influences in the development of the singular noise with which he opposes the noise of contemporary Japanese society. It is never, therefore with Akita, a question of opposing the West and Western influence with traditional Japanese culture—except, perhaps, that element which embraces its own erasure. Whole cities and societies obliterated only to be reborn from nothing, or from debris, or from those real objects that are implacably and silently withdrawn from all the noise of human culture and commerce. Immanent and unconscious, therefore, to the "obliterating white out" that is caused by the ferocious ecstasy of Merzbow is the silence of a creationist sublimation that Lacan finds at the heart of Freud's notion of the death drive.[42] That is to say that the audio unconscious immanent to Merzbow's *a*music seeks obliteration in order to provoke a desire to begin again from the silent rubble in which can be heard the sound of the wilderness.

In his more recent work, Akita has sought to produce a sound that is explicitly posthuman, mobilizing a heterogeneous assemblage of objects. While Merzbow's *a*music seems to refuse any reduction to "themes" or "topics", recent work has been marked by Akita's interest in animal welfare. For example, the theme of the 6CD box set *Houjoue*, the major release of 2006, is the vegetarian history of the early Japanese. "Houjoue means 'ceremony of freeing captured animals'. In the seventh century, the Japanese emperor proclaimed an ordinance prohibiting the eating of animal meat, fish, shellfish, and fowl. In this period, there are many case of *Houjoue* taking place there. I was writing a book called *Cruelty Free Life*, while I was making *Houjoue*. So, both concepts are synchronized".[43] Akita's veganism would seem to sustain a separation between humans, as ethical beings and animals, even as *Houjoue* perhaps evokes the dissolution of that boundary in festival. But while Akita in his statements re-grounds symbolic differentiation between humans and animals in his veganism as a refusal of the wholesale exploitation and utilization of animals, there is no corresponding opposition in Merzbow's *a*music or indeed in his everyday life which he shares with a menagerie of different animals. Furthermore, in his albums Akita has incorporated the sounds of many of his companion creatures most notably birds: bantam chickens, silky chickens, fantail pigeons, ducks, frogs and other animals whose speech, calls, song and responses are recorded and incorporated into noise assemblages. But it is neither easy nor appropriate to try and locate or isolate the sounds of specific species or to suppose their inclusion to indicate something about the content or theme of the work, as if there were some content or theme. There is no point in asking, as Paul Hegarty notes, "where exactly is the chicken? … it is subsumed *Tetsuo*-like into Merzworld".[44] In Merzworld, since there are no words, there is no apparent order of priority.

For Hegarty, Merzworld "is a general ecology of noise" in which there is no clear separation between organic or non-organic life, no clearly distinct human, animal or animal–machine sounds, no categorization of animal life into species, types, forms, anymore than there is a correlative musical organization of sound. Merzbow's work of the mid- to late 2000s is far from unlistenable, often serene and ambient, but it nevertheless evokes a new order that completely revises the imaginary and symbolic distinction which is both predicated upon and supported by a purely dogmatic set of assumptions about the animal kingdom that ground a fundamental opposition between human and animal. In Akita's posthuman ecology, humanity submerges into the crossing of animal and technology.[45] Does this mean that the noise of Merzworld heralds the coming of a new posthuman social order of humananimality in the "wasteful" or aesthetic utilization of technological excess in a complex, no doubt noisy and conflicted incorporation of all creatures and machines in a new order of objects?

The idea of an audio political unconscious, of music as a kind of *wo es war* of political thought is suggested by Jacques Attali when he argues that music, as a particular organization of noise, does not represent but provides a libidinal structure for a future social order.[46] The extremes of Merzbow, however, push at the ecstatic threshold of dis-organization and even obliteration in which there is no difference between noise and music. Any notion of a future social order promised by Merzbow therefore can only be seen as highly equivocal and as precluded as much as pre-empted. But that does not mean that immanent to Merzbow there isn't the possibility of some future thinking of the political and indeed a political unconscious. In Merzbow's *a*music, there is an evocation of a model of the unconscious relative to the assemblage of forces. The "found" ambient music of the world, the music of objects, animals, machines and waste, is not reconfigured by Akita as "voice" in a mirror of human desire and discontent. On the contrary, later works such as *Frog* (2003), *Merzbird* (2004), *Houjoue* (2006), *Bloody Sea* (2006) and *Tumeric* (2006) are conceived as tributes to various living creatures, but do not represent them in any way. The work itself is a participation in noise–music as in a multiplicity of differences, differences of force and intensity of sound. It is participation, a performance of technical know-how, feeling and emotion, thought and intuition in a world of sound in which can be perceived and felt forms of aural life that are an effect of a differential of forces, interior and exterior to every living creature, human or nonhuman. This audio unconscious, comprised of different forces in tension, some louder, suppressing or repressing others, some resisting, returning with greater or a different qualitative force, leaving traces, affecting and being affected and thereby constituted as something or someone. The sounds of objects, even voices, animated by the sonorous force of dissonance, still choiring the music of the clamour for being.

Notes

1. Elisabeth Roudinesco, *Jacques Lacan* (London: Polity Press, 1999), 354.
2. Véronique Voruz, "Psychoanalysis at the Time of the Posthuman: Insisting on the Outside-Sense," *Paragraph* (Special Issue: Psychoanalysis and the Posthuman), eds. Suzanne Dow and Colin Wright (2010): 423–444, 429.
3. Jacques Lacan, *The Seminar of Jacques Lacan Book 1: Freud's Papers on Technique 1954–55* (Cambridge: Cambridge University Press), 319.
4. Colette Soler, *Lacan—The Unconscious Reinvented* (London: Karnac, 2014), 184.

5. See the "object-oriented ontology" of Graham Harman in, for example, *The Quadruple Object* (Winchester: Zero Books, 2011), 121.
6. As Nicola Masciandaro notes, "the primary philosophical sense of *obiectum* is the object of a power," not necessarily a human power, but perhaps an in- or posthuman, power of de-subjectification that is simultaneously the power to objectify. See Nicola Masciandaro, "Afterword" in Eileen Joy et al., *Speculative Medievalisms* (New York: Punctum Press, 2012), 189.
7. Jacques Lacan, *Écrits*, trans. Bruce Fink (New York: Norton, 2006), 197–198.
8. Jacques Lacan, *Encore: The Seminar of Jacques Lacan Book XX: On Feminine Sexuality, The Limits of Love and Knowledge 1972–1973*, ed. Jacques-Alain Miller, trans. Bruce Fink (New York: Norton, 1998), 97.
9. Harman, 40.
10. Jacques Lacan, "Seminar of 21 January 1975," in *Feminine Sexuality*, eds. Juliet Mitchell and Jacqueline Rose (Basingstoke and London: Macmillan, 1982), 167.
11. See Kevin Kelly, *What Technology Wants*. It wants freedom and autonomy, apparently.
12. Brett Woodward, *Merzbook: The Pleasuredome of Noise* (Melbourne: Extreme, 1999), 11.
13. Masami Akita, "The Beauty of Noise: An Interview with Masami Akita of Merzbow," in *Audio Culture*, eds. Christoph Cox and Daniel Warner (London: Continuum, 2008), 59–61, 60.
14. Akita quoted in Woodward, 40.
15. Indeed as Paul Hegarty notes this is characteristic of Merzbow's oeuvre in its entirety, the scope and range of which is impossible to archive.
16. Judith Butler, *Excitable Speech: The Politics of the Performative* (London: Routledge, 1997), 68.
17. Chad Hensley, "The Beauty of Noise: An Interview with Masami Akita of Merzbow," www.esoterra.org/merzbow.
18. Merzbow /Right Brain Audile, *Music for Bondage Performance* (Extreme, 1991).
19. Masami Akita, *'Shitsurakuen': Jobafuku onna hara-kiri (Lost Paradise)* (Right Brain, 1990).
20. Akita, *Music for Bondage Performance*, sleeve notes.
21. Ibid.
22. Akita quoted in Woodward, 26.

23. Ibid., 26.
24. Ibid., 27.
25. Jacques Lacan, *The Ethics of Psychoanalysis: Seminar VII*, trans. Dennis Porter (London: Routledge, 1992), 298.
26. Lacan, *The Ethics of Psychoanalysis*, 149.
27. Akita quoted in Woodward, 27.
28. In the longer version of the "The Beauty of Noise" interview with Chad Hensely that is posted in the Internet, Akita emphasizes that "bondage images are not porn to me," www.esoterra.org/merzbow.
29. Lacan, *The Ethics of Psychoanalysis*, 149.
30. Akita, "Beauty," www.esoterra.org/merzbow. The word for this is "Lolicon" or Lolita-complex which can describe someone with an erotic obsession with under-age girls, or simply the whole style of Japanese anime and manga that is organized around a particular image of an idealized young girl. In a chapter called "A Humbert Humbert for the twenty-first century" from his hikikomori novel *Welcome to the NHK*, Tatsuhiko Takimoto has a lolicon character explain his attraction. The ideal yet wholly fictional image of innocence, purity and femininity embodied in the little girl imaginarily protects consumers from the hostile world of actual social reality. "We're relaxed by the symbol of the little girl. And when they are 2D characters, they have no chance of dealing any blows to our fragile emotional state. On top of that, the motif becomes the weakest character possible in social, physical and emotional senses," Takimoto, *Welcome to the NHK* (Tokyo: Tokyopop, 2002), 76.
31. Harman, 101.
32. Elizabeth Moriarity, "The Communitarian Aspect of Shinto Matsuri," *Asian Folklore Studies* 31, no. 2 (1972): 91–140.
33. Jacques Lacan, "Lituraterre," trans. Jack W. Stone, *Ornicar?* 41 (1987): 5–13, 8.
34. Lacan, "Lituraterre," 5–13, 8.
35. Alexandre Kojève, *Introduction to a Reading of Hegel*, ed. Allan Bloom, trans. James Nicholls, Jr. (Ithaca, NY: Cornell University Press, 1980), 162.
36. Kazushige Shingu, "Freud, Lacan and Japan" (2005), www.discourseunit.com/matrix/shingu.
37. Akita quoted in Woodward, 10.
38. Akita quoted in Katharine Norman, *Sounding Art: Eight Excursions into Electronic Music* (London: Ashgate, 2004), 168.
39. Akita, "The Beauty of Noise," *Cox and Warner*, 61.

40. See Scott Wilson, *Stop Making Sense: Music from the Perspective of the Real* (London: Karnac, 2015), 43, 173.
41. Akita quoted in Hans Ulrich Obrist and Ute Meta Bauer's "Interview with Merzbow," *Nettime* (22 August 1999), www.nettime.org.
42. Norman, 168; Lacan, *The Ethics of Psychoanalysis*, 213.
43. Akita in Roger Batty's "Animal Instincts," *Musique Machine* (4 April 2006), www.musicmachine.com.
44. Paul Hegarty, *Noise/Music: A History* (London: Continuum, 2007), 161–162.
45. Hegarty, 163.
46. Jacques Attali, *Noise: The Political Economy of Music* (Manchester: Manchester University Press, 1979), 4.

11

Melancholy Objects: If Stones Were Lacanian

Timothy Morton

I keep up with research on schizophrenia because my brother Steve has it. (Hence, reading Deleuze and Guattari is always a little strange for me, shall we say.) One hypothesis is that a neurotoxin released by toxoplasmosis gives rise to symptoms of schizophrenia and possibly Alzheimer's. Toxoplasmosis is caused by *Toxoplasma gondii*, a protozoan symbiont that lives in most people's brains harmlessly. Cats are common hosts, and cat poop contains a lot of the protozoans. One theory is that cat poop near to pregnant humans is a way for the symbiont to jump. One in five of us humans carry the symbiont. So when schizophrenic people tell you that aliens have invaded their head and that they need surgery, their paranoia is telling the truth. My brother would talk about aliens and also demand brain surgery "because a piece of my brain has come loose."

T. Morton (✉)
Rice University, Houston, TX, USA
e-mail: tbm2@rice.edu

S. Matviyenko and J. Roof (eds.), *Lacan and the Posthuman*, The Palgrave Lacan Series,
https://doi.org/10.1007/978-3-319-76327-9_11

This hypothesis reminds us that existence is coexistence. Our very bodies are symbiotic communities: our mitochondria, the energy-producing organelles in every cell, have their own DNA, since they are descendants of anaerobic bacteria hiding from the ecological catastrophe they produced, the one called *oxygen*.

What is the phenomenology of symbiosis? How does it arise? Symbiosis seems to have contours similar to melancholia. Melancholy is an object-like presence that our psyche finds hard to digest. It is literally the footprint of another entity of whatever kind whose proximity was experienced as a trauma (the Freudian logic of the death drive).[1] Melancholy by definition implies coexistence, which is why it is a significant effect for thinking ecology, since ecology is about coexistence thought as widely and as deeply as possible. This coexistence need not be with sentient beings, nor even with life forms per se: it can include all entities ("objects" in the terminology I shall be using here) such as rocks, plutonium and carbon dioxide. Mourning is that attunement in which one entity is digested by another, and then, the very footprint, the trace of that entity, is also digested. For mourning to exist, melancholy must exist as its condition of possibility. Thinking ecology thus is not separated from the probes sent from other entities, the probes we find embedded in us as a condition of existence.

But just as importantly, melancholy does not imply anything about subjectivity. All you need for melancholy are various kinds of objects. This is what makes it different, in traditional psychoanalytic theories, from other effects. Indeed, melancholy speaks a truth of all objects—I use the term "object" in a value-neutral way, implying any real entity whatsoever, not objectification or subject–object dualism: a person, a blade of grass, a crystal, a planet. In "Experience," Emerson writes, "I take this evanescence and lubricity of all objects, which lets them slip through our fingers then when we clutch hardest, to be the most unhandsome part of our condition."[2] From there, Cavell develops what he calls the "standoffishness" of objects. But melancholy doesn't require fully formed subjectivity—indeed, subjectivity is a result of an abnegation of the melancholic abject as Kristeva argues, so it's a positive hindrance in this sense.[3] So we can generalize Cavell so that all objects are standoffish with regard to one another, not just to humans.

The rather beautifully brittle quasi-animistic dandyism (what a combination!) of calling objects "standoffish" is an index of something incapable of being *handled*—both in the physical and psychological senses. This "something" is what this essay will now call *the symbiotic real*, the biospheric whole of which this thing here, this thing doing the typing of this sentence, is a part, in some uneasy, non-theistic holist sense, a sense that doesn't imply that wholes are greater than the sums of their parts. This is because the parts always outnumber the whole, and since whole and parts exist in the same way, a whole is always a not-all set, at least one of whose members slips away from at least one other (in this case, an entity and the thinking human trying to access that entity) like the soap in the bath. The concepts that reify or cover over the symbiotic real, such as that old chestnut, Nature, tend to contain something like this fact as a symptom. When one looks for Nature, for example, one ends up finding trees, frogs, clouds and highways. But Nature promises to be something both physically and ontologically bigger than frogs and clouds and more embracing than trees and highways. It becomes reassuring that one can't find it when one looks, rather than unsettling.

This means that the appropriate philosophy for an ecological era is an *object-oriented ontology* (OOO) that respects the withdrawn strangeness of objects while simultaneously: (1) not discriminating against them in any way (reductionism, holism, anthropocentrism, biocentrism) and (2) allowing for their uncompromising unicity, the fact that they obtrude upon ourselves and upon one another through time immemorial and the vastness of space.

OOO is deeply congruent with what ecology, since the ecological age is what this essay now calls *the time of hyperobjects*: the moment in which human beings realize that they are enmeshed in a series of entities such as climate (and global warming) and evolution (and the plenum of life forms) that are massively distributed in time and space in such a way that human access modes (including algorithms running on nano-speed timing) are incapable of representing fully, for instance "in real time."[4] From here, we can draw up a new ethical map that is neither utilitarian nor holistic—both systems fail in the face of hyperobjects. It is nonhuman (and even nonliving) entities that show us the way towards this new ethics. By "standing in the place of the death drive," as psychoanalysis would put it, humans have the chance to coexist non-violently with other beings. Indeed, this traumatic coexistence is a better reading of the Freudian death drive than one that presents it as a simple existential human drama.

What Melancholy Says About Ontology

This essay will now begin an investigation of melancholy that leads to some facts about things in general: a lump of plutonium, a village, an ashtray.

Mourning is just the collapse of an object into its *appearance*, namely its appearance for some other entity. When I die, I become your memories, some paper crumpled in a waste-paper basket. When I fall into a black hole, you see an image of my horrified face on the surface of the event horizon, an image that slowly but surely fades. Appearance is a photograph of the past. Appearance is form. Form is the past. When we study the form of anything—a poem, some rock strata, a piece of software, a pyramid—we are archaeologists studying the past.

Melancholy, however, is the holding open of a *rift* between the *appearance* of a thing and its *essence*. This rift cannot be located anywhere in given, ontic space-time. I look over the surface of a thing and do not find it—I probe its depths and also fail to find it. Indeed, this probing might double or further multiply the problem: when I split a piece of chalk to find out what it is, I now have two pieces of chalk.[5] This is because the essence of a thing is not the thing "itself" devoid of aesthetic appearances, but rather the withdrawn reality of a thing. We can never get our hands on a thing, only appearances: our holding the hammer is not a hammer; our breaking the hammer is not a hammer; our thousand-word poem about the hammer is not the hammer. Furthermore, and this is the key point: *even the hammer is not the hammer*. The hammer is different from itself, already. Even if the hammer developed Muppet-like powers of speech, it would not be able to grasp its essence.

As we just saw, Emerson gives voice to something Kant had thought yet suppressed in his own way, situating the Special Thing that does the tying of loose ends not in some external Nature but in the operations of the (human) mind, a part of me that withdraws just as Nature does when I try to point to it. Let's visit that line again: "I take this evanescence and lubricity of all objects, which lets them slip through our fingers then when we clutch hardest, to be the most unhandsome part of our condition."[6] In so doing, Emerson focuses on the human, but we can begin to generalize his remark. Things in general elude themselves and one another. They are lubricious, like soap in the bathtub.

Existence is coexistence, and coexistence is melancholy. And melancholy just is the persistence of a *rift* between essence and appearance. Is this not what gives Lars von Trier's film *Melancholia* its stunning appeal?[7] And is this not the reason why it was a brilliant choice to use the prelude of Wagner's *Tristan and Isolde*, with its agonizingly beautiful "Tristan chord," a chord that can't make up its mind, a chord that embodies the rift under discussion here? The chord is suspended, just as the rift suspends. From this suspension flows what we call time and space. Time is not a neutral box in which objects float, but is rather an emission of objects themselves, part of their appearance, the way in which they are always appearing slightly different from their own selves. Likewise, space. Happily, this is quite in accord with Einstein, for whom space-time is an emergent property of objects, not a rigid framework in which objects are extended.[8]

Thus also the choice to depict people, birds and other things moving in ultra-slow motion, like the video art of Bill Viola in the opening shot of *Melancholia*, was also a stroke of genius. A stroke of genius, since slow motion evokes the way in which time seeps out of things themselves: time and space are aesthetic phenomena. Indeed, there is some kind of gesture towards scientific realism here, since the proximity of a massive object's gravitational field would indeed slow down time, and this is the plot of the film: a gigantic planet, a hyperobject par excellence, is heading on a collision course with Earth. It is as if everything in that opening sequence is happening in the presence of a gigantic entity that is invisible, rendered as slow motion. As if we, outside the movie screen, existed in one temporality that was suddenly seen as relative, not absolute, distorted by awareness of a huge object. The invisible withdrawn essence of the thing is made visible in the slow motion and in the agonizing, wonderful-horrible chords of *Tristan and Isolde*, the opera that confuses love and death.

In the Shadow of a Giant Object

Melancholia perhaps makes another reference, namely to Zeno's paradox. Zeno refutes the idea that space and time can be subdivided, by imagining motion as a sequence of stillnesses that can be divided infinitely, thus never arriving at its destination. Cinema, too, imagines movement as a sequence of stillnesses, or stills. Time ceases to be a container in which things float. We see something like what Deleuze calls a *time crystal*, some kind of higher dimensional being made up of the flowing filmstock and the imagery inscribed on it. The stroke of genius was for von Trier to introduce the film with this crystal, rather than have it emerge as if organically from the flow of the film. As if he were trying to convey, within a fluid medium, a horrifying stasis, a Parmenidean One showing through the fluid. A singular, unrepeatable unit: the moment of death. The inevitability of there being at least one (other) being out there. As Justine, the protagonist, puts it, "I know things. I know that we're alone." What does she know? That there is a reality beyond the human. She feels it in her bones.

The ultra-slow sequence that opens the film depicts the last minutes of life on Earth. The knowledge of the end, of closure, which is not simply a mathematical end but rather the *feeling* of ending, that is, the collapse of the rift between essence and appearance, leaving behind only the shadow of appearance as such, becomes the very object of cinephile delight, our agonized pleasurable inspection. It is as if closure is fed back into the feeling of *being in the middle*—that is the feeling of existence, of things being suspended, the rifting of the rift from which flow the rippling waves of space-time. And that this feeling of middle is itself fed into *aperture*, the feeling of beginning, which is a sublime interruption of a smooth ignorance, a double-take in which we always belatedly notice the existence of a fresh rift in things, a fresh rift between essence and appearance. *We know how this will end*. The open, dark futurality of a thing is suspended in slow motion before our eyes.

An action replay—this is yet another reference in that opening shot. Something has always already happened—this defines beginning. Lovingly, melancholically, fetishistically, an action replay relives a moment we already know is coming. Are we digesting life into inert inorganic "quiescence" (as James Strachey beautifully translates the Freud) or are we being stirred into yet more agonizing life?[9] The sadistic accuracy of sports television and slow-motion orgasm shots in pornography becomes the masochistic intensity of reliving a moment in which we know we are not outside the universe looking in, like spectators in a cinema, but are wedged firmly in between other beings, human, sentient, living, nonliving. The film gazes at us. The film is an analogue for the gigantic planet, coming asymptotically ever closer. A rift, whose very existence is the premonition of another death, thus speaks to a radical finitude of being.

Form is appearance. *Appearance is the past.* Another way to say this is that appearance is the footprint of at least one (other) object. This glass is a record of how a hot liquid was poured into a mould, how my greasy fingers slipped one day and chipped it with a knife, how the dishwasher dulled its sparkle. A poem is a certain form: just this lineation, just that rhyme scheme, just this stanza form, just those images. Poems are records of causal-aesthetic decisions. To read a poem is to be an archaeologist. For OOO, the physical form of an object is a form-as and a formed-by. Freud argues that the ego is just the record of "abandoned object cathexes."[10] What if we inverted this phrase and assert that the form of objects is their ego?

A Rift in Being

If ego is object-like, then the inverse applies. The identity of this glass is the way it was shaped as a glass. Form is *memory*, as in a memory stick: your face, your hard drive, your chipped coffee mug, records what happened to it. What is called the past is really other objects that coexist with the object in question. When we hold a glass, we are holding the past. As we saw earlier, there is a profound *rift* between the *appearance* of the glass and the *essence* of the glass, which is not the same as the difference between an undifferentiated blob and a defined shape with stem, neck, weight, sparkle and so on. For lack of a better way of putting it, it's *the difference between the glass and the glass.* ("What is the difference between a duck? One of its legs is both the same.") The glass is a glass and an uncanny not-glass.

What we are dealing with here is a Lacanian form of set theory that allows for the existence of the dreaded Russell Set Paradox: the set of things that are not members of themselves. The lopsided plentitude of the symbiotic real seems to demand a weird set theory that circumvents Russell's norms about what a legal immigrant into the land of sets would look like. A weird set theory allows for beings to be themselves and not-themselves simultaneously, indeed as the very condition of their possibility. Such lying, self-contradictory beings seem to be illogical, lying beyond the law of noncontradiction. But as Lacan himself observed, *there is no metalanguage*. The concept of metalanguage was developed by the logician Tarski precisely to ward off the threat of Liars: sentences such as *This sentence is false.* The sentence is both true and false at the same time. If you have a metalanguage that can specify in advance what a sentence is, then you can set heavy bouncers at the doorway of the nightclub of logic. The trouble is, our Liar can get around the bouncers. If for instance, *This sentence is false* is not a sentence, then what about this:

This is not a sentence.

For every bouncer, one can find a proposition that slips through. The insight that there is no metalanguage is Lacan's wonderfully boiled down version of Heideggerian and Husserlian phenomenology, which argues that what we think and how we are thinking it—and indeed things in the world we are thinking about—are locked together in such a fashion that we are strictly unable to peel ourselves away from them, or peel them apart, without some fundamental alteration taking place. Seeing a thing, in other words, is never from some extra-real vantage point of safety, a VIP lounge in the nightclub of being.[11]

Tarski and Russell ward off the infiltrators into the nightclub of being as part of a sustained reaction against Cantor's rather horrifying discovery of the abyssal depth of reason. Infinity, as Kant argued, could be thought even as it could not be understood. It could even be felt, in the aesthetics of the sublime.[12] Cantor's deliciously simply diagonal proof discovered a series of numbers beyond the infinite set of rational numbers, a set that could not be contained within that set. Indeed, the set of real numbers contained the set of rational numbers, although both were infinite—so that the infinite set of real numbers is, absurdly yet visibly (you can see it on the page in the diagonal proof), larger than the other. Furthermore, there are no smooth continua, no bridges, between the larger and the smaller infinity. We have a set whose members are not members. An infinity contained within another infinity to which it cannot add up. Cantor saw further into the abyss of reason than Kant had done, and philosophy was confronted with the possibility that logic might not totally exclude contradictions. Either such things are outside of logic, or logic is brittle in some way when it polices contradiction, a rule in Aristotle's *Metaphysics* that is very widely accepted, but which has not been proved. Perhaps the law of noncontradiction is a sort of "observation selection effect" based on the human position within reality, such that if we want to formulate a truly posthuman thought, thinking must go beyond it.[13] What this essay argues in particular is that the reason why true lies, dialetheias (Greek, double-truth), exist, is that something about beings in themselves is contradictory at a fundamental level. If we really want to escape a world in which there are duplicitous, reflexive humans and bland everything else, we might want to entertain this idea.

Now, we shall return to the argument that appearance is the past. The existence of a form is the existence of a kind of memory, which is evidence of a withdrawn essence. Freud argues that the unconscious is some kind of inscribable surface. He uses the analogy of the mystic writing pad. Derrida has a marvellous, McLuhan-like essay on it ("Freud and the Scene of Writing") since Freud is in effect admitting that the unconscious is what Derrida calls *arche-writing*, namely a technological device that subtends meaning.[14] When you use a mystic writing pad, you erase the wax paper, but the impression of the writing stays on the wax tablet beneath. Script is inscribed in an object. What Freud is hinting at is that memory is object-like and that objects have memory.

Derrida allows for contradictory things to some extent—linguistic systems are contradictory, because they must, as a condition of their existence, both include and exclude the technological device, tool, surface (all synonyms) in which or on which or via which they are inscribed. What this essay is claiming, in a sense, is that Derrida spies a tiny region of a much larger space of beings, a region populated by humans and their writing tools—a region that is necessarily contaminated with nonhumans (ink, typewriters, paper, computers, language itself). What this essay argues is that Derrida descries the rift between what is here called *appearance* and *essence*. These delightfully antique words have become contaminated by the beliefs we have about them, so that we think appearances are candy sprinkles on top of the bland cupcake of essence. This default Aristotelianism is the case even within the most sophisticated interpretations of science. This default view assumes that objects are real to the extent that they are "constantly present," as Heidegger puts it.[15] Extending Heidegger's assault on this metaphysics of presence, Derrida shows how meaning systems are inherently dislocated, always already. Within this spider-webbed corner of hermeneutics, we can discern an exit hole towards ontology, despite the taboos against ontology that Derrida himself would have reminded me of if he had just read the essay up to this point. Let us then try to see if we might elaborate *essence* without the metaphysics of presence.[16]

Strung Out

Kant was the first in the West to step outside of the metaphysics of presence, of givenness. I never have the essence of a thing, only its appearance: with this basic insight, Kant reduced prior metaphysics to vulnerable islands floating on top of a gigantic ocean of reason, a third and unexplored dimension of thinking. The reasons why things happen (causality) are never directly available to me: what I see—science is very Humean on this point, and Kant fills in the reasons for Hume's argument—are clouds of data that correlate statistically.[17] The very appearance of a thing is a melancholy trace of its essence, which I am unable to locate anywhere in ontically given space or time. No point on the surface of the glass corresponds to its essence as a glass, no matter how high-dimensional we make this surface. Nowhere in a 4-D glass, for instance, a rather strange-looking worm that included the first moulding of the glass and its final shattering, would we find the location of its essence. *Surface* here means anywhere at all that can be located and measured: I can turn the glass inside out in my thought, I can send probes that penetrate its crystal lattices—nowhere will I find its essence.

This withdrawn essence is nowhere, but in this glass I am holding. This is a glass, not a coyote. Yet it is unavailable to me. Moreover, the essence of the glass is unavailable to any other entity whatsoever. These other entities include the very things that formed the glass—mould, molten sand, dishwasher, knife, greasy fingers. My melancholy experience of the glass by definition is only one of a vast host of melancholy apprehensions of the glass, in whatever way moulds, sand, dishwashers and so on carry out their business. I am far from saying that my dishwasher is alive or sentient. I am saying in fact that my sentience and aliveness are not very different from the appearance of a dishwasher. When, in a state of depression, I feel like an inert lump, a stone or a fallen log, infested with thousands of crawling insects as I lie helpless on the forest floor, my own time oozing past too slowly to deal with the thousand prickling beetles on my surface, I am being a realist, I am glimpsing something true about my being.

There is thus a profound truth in the idea that depression is a kind of frozen wisdom, a philosophical insight expressing itself as a slow-motion scream or inertia. Inertia just is the persistence of a rift between essence and appearance. Theories of motion that imagine objects to be self-identical, consistent things that move from A to B "in" time, run into the Zeno's paradox phenomenon we explored earlier. But if an entity is riven from the inside "between itself"—we must stretch grammar to breaking point here—then it has all the resources it needs to move and change, and in particular we can explain inertia, which just is how an object keeps going, or persisting, without external input.

What is frozen inside melancholia? The profound philosophical insight that a thing, in order to exist, is radically open. What is existing, or continuing, or persisting? It just means being in-difference from oneself. Existing thus is futural. It is not-yet, strung out—Heidegger's exploration of Angst shows how this strung-out quality is intrinsic to at least one being, the human.[18] "Strung out" does not mean "stretched across *n* now-points": in that case, we would have a metaphysics of presence concerning time. Time would itself be a sequence of atomic dots, or some other pregiven thing—and pregivenness is what is precisely at issue, since Kant.

Thus, the essence of a thing is not locatable on its ontically given surface, no matter how many dimensions we bestow on this surface. The "depth" of a thing is not some point in 3-D space, so the essence of the glass is not measurably "inside" its slender ballooning concave. I won't find its essence anywhere in a 4-D space either that allows for different moments of the glass's existence. The emergence, existence and destruction of the glass are equal aspects of its glassness.

Literary scholars have something on hand that will greatly help to clarify this point. Consider a poem. Its meaning is its future. At some point, we will read it and decide on its meaning. Then, we re-read it and another meaning might emerge. The only reason we return to a poem is that it might release a different meaning this time. Since the aesthetic dimension is the causal dimension, what does this basic fact about what we do as literary scholars tell us about time itself? It tells us that the "present" is not a bubble between past and future, or a blinking cursor, or a point. The present is a construct imposed on an uncanny intermeshing of appearance and essence. Presence is hollowed out from the inside by "past" and "future." Readers know that time is not a bland container of things and events—they simply mistrust this basic insight, relegating it to a small island of (human) subjectivity in a sea of scientism.

The meaning of a poem is (in) the future. This future is not a now-point that is *n* now-points away from the current one: it is withdrawn, it is withdrawal. This future is what Derrida calls *l'avenir*, the to-come, or what I call the *future future*.[19] In a strict sense, poetry does come from the future, just as Shelley argues. If Aristotle rules the past, Plato seems to be in charge of the future, in the ontology we are sketching out here. A weird Platonism is in effect, beaming the shadows of objects down from the future future into sensual-aesthetic-causal coexistence. The future future is not some transcendental beyond: this would be a top object par excellence. Nor is the future future a "time in which" the object "resides." The future future is the pure possibility of the object as such. *The essence of a thing is the future.*

Withdrawal is futurality, not as a predictable time that is ontically given. Nor is futurality a poststructuralist "excess," since this implies a thing for which the object is excessive (this could be a telescope or a teabag as much as it could be a human or a fish). Excess is an appearance, belonging to the realm of an object's pastness. Nor is futurality a void, a gap or empty nothingness. Perhaps the term *openness* expresses it best. Withdrawal is openness. Now, we can discern more clearly the rift (Greek, *chōrismos*) between essence and appearance. It is a rift between openness and pretence. An object persists and moves for as long as it can maintain its inner lie. If it is forced to speak nothing but the truth, destruction ensues: the rift collapses.

The Realm of Death

Thus, coexistence is inherently non-violent: it tolerates the rifting of as many things as possible, which means that it lets them exist. A thing is fragile, not because it can be destroyed—but because the condition of this possibility of destruction is the very rift between essence and appearance. Destruction just is when something resonates with the inner wound, the *hamartia*, of a thing, whether that thing is Hamlet or a wine glass. When an opera singer sings just the right note attuned to the resonant frequency of the glass, it ripples as if it were having an orgasm or a stroke and then bursts into fragments. New entities are born, shards of glass, and the destroyed glass is nowhere to be seen, just as my own death is nowhere to be found in ontic, given space.

A rift is an aggravating thing, like that dissonant Tristan chord. A thing is fragile, unstable. This fragility seems at least to be a condition for life. DNA is a molecule that is trying to cancel itself out: some inner disequilibrium moves it to erase the stain of itself from things. But in so doing, it repeats itself. Don't we see here, quite literally, the fact that death comes before life? If you think that DNA is alive, then you have stretched the definition of "living" to include things such as computer viruses and artificial life. For a start, what distinguishes a life form from non-life? Not very much. Life forms are made of non-life. For another thing, doesn't life consist in a frantic effort to get rid of the stain of itself, all the way down to the DNA level—and indeed beyond, since RNA and other replicators predate DNA? What is called *life* just is the expression of unstable molecules that contain some kind of inner disequilibrium.

In their attempt to cancel themselves out, to solve their inner paradox, they ironically produce more of themselves. The frantic repetition of life is what Freud calls *death drive*, for this very reason. The effort to kill the zombie reproduces the zombie. Freud's only mistake was not to go far enough—to imply, though not to demonstrate, that death comes from beyond life, that life is, as Nietzsche might put it, a rare form of death. Freud stops at the limit of a single-celled organism in his exploration of the death drive.[20] But there is no good reason to stop there.

What if the reason DNA does this is because other nonliving beings also do it? Silicate crystals can also self-replicate. Before DNA, there was RNA, which hitched a ride on replicant silicate crystals. When I break the glass, a remote descendant of that ancient silicon, a new beginning occurs: the rift is multiplied into tens of glassy fragments. The logic of the death drive appears to be default to existence as such, beyond life. Freud examines the manifestations of the death drive down to the scale of a single-celled organism in *Beyond the Pleasure Principle*. But surely we can go at least as far as DNA, which itself is an uncanny kluge of viral code insertions, viroids, DNA code from thousands of other species mixed together, "junk" and so on. And why stop there? DNA requires ribosomes and ribosomes require DNA, so there is a vicious circle. To break it, there must have been an RNA world in which DNA had not yet developed. In RNA world, RNA molecules attached to some other replicator, perhaps a silicate crystal of some kind. So now we have uncanny repetition at the level of non-life, just like the puppet Olympia in *The Sandman*. Life is uncanny, because it is both itself and not-itself at once: a basic repetition throbs within its inner structure.[21]

If this is the case, we ought not to stop even there. If it's the case already that RNA, a nonsentient molecule (nonsentient at least in the standard definition of sentience), exhibits some kind of "drive" beyond "life" do away with itself, then surely other nonsentient beings might be the same. Surely there is a reason why replicating molecules can exist in the first place. And if, to solve its inner paradox, RNA hitched a ride on a silicate that was also a replicator, why confine our search for the death drive to organic molecules? Why can a molecule such as a self-replicating silicate crystal arise in the first place? It seems likely that such replicators are themselves expressions of an inherent instability in objects in general. Indeed, this instability might not be confined to the "building blocks" out of which things are made, but might affect "macro" scale entities such as the badgers and bottle-nosed dolphins made out of DNA's attempt to cancel itself out. And what about other things?

Objects are lame, they *halt*, like Turing's machines—objects limp along, hamstrung by their inner inconsistency.[22] In the movie *Plastic Bag*, a designed thing, the eponymous bag, its inner monologue narrated by Werner Herzog (who else?), floats around in the ocean near Midway Atoll, the huge vortex of spinning bags in the Atlantic. It wishes that it could ask just one question of its maker: "If I could meet my maker, I would tell her just one thing: I wish that she had created me so that I could die."[23] Do all things yearn for consistency, for death? It is certainly the ecological emergency that has shown us objects that obscenely outlive their makers: a Styrofoam cup, plutonium, global warming, a plastic bag. These are the beings that, like Tithonus condemned to eternal aging, force us to see the vast, ungainly objects in which and with which we have our being.

Standing in the place of the death drive, then, the ultimate ethical position that Lacan sketches out in his study of the still uncanny, still terrifying Sophocles play *Antigone*, is precisely being as *weak* as possible. Allowing oneself, in other words, to coexist with as many lame beings as one possibly can, standing in their midst, in the midst of the death drive, the insatiable quest to cancel itself out, the frenzy that makes more of itself in the very attempt to erase itself. This frenzy takes place beyond life, since it is life's condition of possibility. This universe of unique objects, riven from within between essence and appearance, all uniquely lame (uniquely wounded), is not so much an animist universe as a *re-animist* one, a zombie universe of undeath. A standoffish approach to "handling" such a universe, such as the perennial post-Kantian goth culture, seems to be in the right key signature, and so we return to the truth within the noise that Cavell's dandyish concept of the standoffishness of objects is making.

I do not propose a vitalism then, so much as an *undeathism*. In this universe of objects condemned to undeath by their own inner inconsistency, until some magic bullet hits them and they crumple into pure appearance, we can stand like a willow tree. Antigone's lover Haimon pleads to his father Creon to bend like a flexible tree in a violent storm: "Seest thou, beside the wintry torrent's course, how the trees that yield to it save every twig, while the stiff-necked perish root and branch?"[24] Trying to get rid of ecologically induced depression could be a violence that risks reproducing the very conditions that imperil coexistence. Instead, we might listen to what the depression is telling us.

Notes

1. Sigmund Freud, *On Murder, Mourning and Melancholia* (London: Penguin, 2007), 201–218.
2. Stanley Cavell, *This New Yet Unapproachable America: Lectures After Emerson After Wittgenstein* (Albuquerque: Living Batch Press, 1989), 86–88; Ralph Waldo Emerson, "Experience," *Essential Writings*, eds. Brooks Atkinson and Mary Oliver (Modern Library, 2000), 307–326 (309).
3. Julia Kristeva, *Powers of Horror: An Essay on Abjection*, trans. Leon S. Roudiez (New York: Columbia University Press, 1982).
4. See for example Timothy Morton, *The Ecological Thought* (Cambridge: Harvard University Press, 2010), 130–135.
5. Martin Heidegger, *What Is a Thing?*, trans. W.B. Barton and Vera Deutsch, analysis by Eugene T. Gendlin (Chicago: Henry Regnery, 1967), 19.
6. Emerson, "Experience," 309.
7. *Melancholia*, directed by Lars von Trier (Nordisk Film, 2011).
8. Albert Einstein, *Relativity: The Special and the General Theory* (London: Penguin, 2006).
9. Sigmund Freud, *Beyond the Pleasure Principle*, ed. and trans. James Strachey (New York: Liveright, 1950), 86.
10. Sigmund Freud, *The Ego and the Id*, revised and ed. James Strachey, intro. Peter Gay, trans. Joan Riviere (New York: Norton, 1989), 24.
11. Graham Priest, *In Contradiction: A Study of the Transconsistent* (Oxford: Oxford University Press, 2006), 10–12.

12. Immanuel Kant, *Critique of Judgment: Including the First Introduction*, trans. Werner Pluhar (Indianapolis: Hackett, 1987), 103–117.
13. Nick Bostrom, *Anthropic Bias: Observation Selection Effects in Science and Philosophy* (New York: Routledge, 2002).
14. Jacques Derrida, "Freud and the Scene of Writing," *Writing and Difference*, trans. Alan Bass (London and Henley: Routledge and Kegan Paul, 1978), 196–231.
15. Martin Heidegger, *Being and Time*, trans. Joan Stambaugh (Albany, N.Y: State University of New York Press, 1996), 89, 90, 92, 108.
16. Readers familiar with movements with speculative realism will note that this is an alternative way of thinking the problems laid out by Kant from which we have suffered ever since, an alternative that is to Quentin Meillassoux. Meillassoux gets rid of the human–world correlate and preserves the law of noncontradiction. He thus allows for radical contingency—anything at all could happen. This essay simply multiplies the correlate everywhere, so that it becomes the glass–table correlate, the iPhone–wallet correlate and the Melancholia–Earth correlate. At the same time, it violates the law of noncontradiction, allowing for some beings to be self-contradictory.
17. Judea Pearl, *Causality: Models, Reasoning, and Inference* (Cambridge: Cambridge University Press, 2010).
18. Heidegger, *Being and Time*, 299, 309–311.
19. Jacques Derrida, "Hostipitality," trans. Barry Stocker with Forbes Matlock, *Angelaki* 5, no. 3 (December, 2000), 3–18.
20. Freud, *Beyond the Pleasure Principle.*
21. Sigmund Freud, *The Uncanny* (London: Penguin, 2003).
22. *Oxford English Dictionary*, "halt," v1.1., accessed September 29, 2011, http://www.oed.com.
23. *Plastic Bag*, directed by Ramin Bahrani (Noruz Films and Gigantic Pictures, 2009).
24. Haemon in *Antigone*, trans. R.C. Jebb, http://classics.mit.edu/Sophocles/antigone.html.

Index

S. Matviyenko and J. Roof (eds.), *Lacan and the Posthuman*, The Palgrave Lacan Series,
https://doi.org/10.1007/978-3-319-76327-9

MIX

Papier | Fördert gute Waldnutzung

FSC® C083411

Kolibri 360 GmbH
Ferdinand-Jühlke-Straße 7
99095 Erfurt, Deutschland
produktsicherheit@kolibri360.de